Linking Research
to Crop Production

Linking Research to Crop Production

Edited by
Richard C. Staples
Boyce Thompson Institute for Plant Research
Ithaca, New York
and
Ronald J. Kuhr
Cornell University
Ithaca, New York

Springer Science+Business Media, LLC

Library of Congress Cataloging in Publication Data

Boyce Thompson Institute for Plant Research Conference on Linking Basic Research
to Crop Improvement Programs for Less Developed Countries, Cornell University,
1979.
Liking research to crop production.

Includes index.
1. Agricultural research — Congresses. 1. Underdeveloped areas — Plants, Cultivated —
Congresses. 3. Underdeveloped areas — Agriculture — Congresses. I. Staples, Richard C.
II. Kuhr, Ronald J. III. Boyce Thompson Institute for Plant Research, Yonkers,
N.Y. IV. Title.
SB51.B68 1979   631   79-25737
ISBN 978-1-4684-1023-5   ISBN 978-1-4684-1021-1 (eBook)
DOI 10.1007/978-1-4684-1021-1

Proceedings of the Boyce Thompson Institute for Plant Research Conference
on Linking Basic Research to Crop Improvement Programs for Less Developed
Countries, held at Cornell University, Ithaca, New York, April 25, 1979.

© 1980 Springer Science+Business Media New York
   Originally published by Plenum Press, New York in 1980
Softcover reprint of the hardcover 1st edition 1980

Acknowledgments

The conference from which the papers in this book were drawn was held at Boyce Thompson Institute for Plant Research on the campus of Cornell University on April 25 and 26, 1979. This conference was held to mark the dedication of the new Institute facilities and was made possible by generous contributions from the Ford Foundation, Cornell University College of Agriculture and Life Sciences and the Institute.

Preparation of the manuscript would not have been possible without skillful typing by Joanne Martin and thoughtful preparation of the index by Kathryn W. Torgeson.

Preface

In 1941, The Rockefeller Foundation sent a team of three American agricultural scientists to Mexico to survey the prospects for increasing grain production there. The nature of the program that was subsequently established by the Ford and Rockefeller Foundations has had a large influence upon the evolution of agricultural research for the developing countries, and the project grew into what is now called the Green Revolution.

The Green Revolution has been vastly successful because there was abundant research and technology available to draw upon. Now the Green Revolution has evolved into a very complex program of development; the momentum has slowed; and it appears that the time has come to reconstruct the research base which underlies the cropping systems for the third world.

What are some of the problems that we face? The expanding world population is taking up more living space just when land is urgently needed to feed the 6.3 billion persons projected for the year 2000. The causes of the population problem are deeply imprinted in the social pattern of most countries, and certainly there are no simple solutions in a nation unwilling to restrain population growth. The problem of population growth, and others like it, is very much a sociological problem.

Unemployment also poses a serious problem. The world's poor still exist on marginal diets. India, for example, now exports bumper crops every year, but thousands still beg in the streets of Calcutta. It seems that the social, economic, and political problems which attend development are now as urgent as the problem of food production.

Somehow solutions to these problems of population growth and unemployment need to be made a part of our research goals. Research for development should not be concerned solely with crops, but should also involve a consideration of such problems

as climate, urban employment, education, transportation, population control and market incentives. Concerned as we are here
with research for crop production, can the goals that we develop
be refined to include a better synthesis of the agricultural,
social, political and economic sciences?

In part, however, the Green Revolution has slowed down because
we are facing tougher production problems. Progress in developing
crop varieties which are more tolerant to adverse environmental
conditions -- cold temperature, poor soils, heat and drought --
has been slower than expected, although there have been substantial
gains in developing pest- and disease-resistant crops.

Given these difficulties, how much can we hope to accomplish?
No one knows what the upper limits of productivity really are,
but we may be approaching a yield plateau for some crops. USDA
statistics published annually have shown that while productivity
per hectare for soybeans and corn is still rising, productivity
for rice and cotton is topping out. Eventually we may need to
turn to several alternative crops, such as the winged bean, if
sociological factors will allow it. This crop and several others
might provide more room for improvement than such finely tuned
crops as corn.

The technique of genetic engineering for crop improvement
might be an especially useful tool to bypass some of the genetic
mechanisms which appear to hinder rapid progress in crop improvement. Nevertheless, some aspects of genetic engineering, including
the transformation of cells from crop plants using bacterial
plasmids, seem a long way off. Certainly there seems to be no
reason yet to avoid a fuller evaluation of the earth's germplasm collection, since often less than 20% of most crop collections has been tested.

Obviously, then, crop productivity is only part of a broader,
more complex problem of development of appropriate technology
How can we, as scientists, contribute best to this problem
which is of increasing importance to those of us who live in
the first world, far removed from the sites of application?
How should we export our knowledge without destabilizing the
societies we want to help? These are some of the questions
that should be raised here in order to better link our efforts
to those in that poorest third of the world.

Richard C. Staples
Ronald J. Kuhr

Ithaca, New York
July 1, 1979

Contents

LINKING BASIC RESEARCH TO CROP IMPROVEMENT PROGRAMS
IN DEVELOPING COUNTRIES -- Ruben L. Villareal   1

    Agricultural Research in Developing Countries
    Application of Basic Research in Crop
      Improvement
    Areas of Greater Value to Developing Countries

THE EVALUATION AND REMOVAL OF CONSTRAINTS TO CROP
PRODUCTION -- N.C. Brady   11

    Unique Significance of Asia and of Rice
    Rice as the Food Staple of the Poor
    Three Sources of More Food
    The Potential Performing Gap
    Constraints on Rice Yields in the Tropics
    Research to Remove Constraints
    Research to Permit Tolerance of the Stresses
    Research to Support the "Factory"
    Symbiotic Alliances

CROP IMPROVEMENT -- John K. Coulter   35

    Introduction
    The Need for Improved Technology
    Technology Development
    Conservation and Utilization of Genetic Resources
    Crop Improvement
    Objectives in Crop Improvement at the Centers
    Pest Resistance
    Breeding for Adverse Environments
    Plant Quarantine
    Basic Research
    Conclusions

THE ROLE OF PHYSIOLOGY IN CROP IMPROVEMENT -- Paul J.
   Kramer   51

      Introduction
      Limitations on the Contributions of Plant
         Physiology
      Photosynthesis and Yield
      Yield Usually is Limited by Environmental
         Factors
      How to Increase the Contributions of Plant
         Physiology
      General Discussion
      Summary

BLUE ROSES AND BLACK TULIPS: IS THE NEW PLANT
   GENETICS ONLY ORNAMENTAL? -- Peter S. Carlson   63

      Introduction
      Some of the Components of the New Genetics
      Completely Defined Cellular Systems are Not
         Required for Use in Crop Improvement
      What Kind of Genetic Variability is
         Agronomically Desirable?
      Envoi

BIOMASS PRODUCTION AND UTILIZATION -- H.T. Huang and
   L.G. Mayfield   79

      Nature of Linkage and Consequences
      Funding Allocation for Agricultural R and D
      Biomass Conversion
      Arid Land Plants

BASIC RESEARCH IN BIOMASS PRODUCTION: SCIENTIFIC
   OPPORTUNITIES AND ORGANIZATIONAL CHALLENGES --
   Israel Zelitch   101

      Basic Research Needs
      Relation of Net Photosynthesis and Photo-
         respiration to Crop Yield
      The Glycolate Pathway of Photorespiration
         and its Regulation
      Producing Mutants of Higher Plants by Selections
         on Plant Cells
      Economic Benefits of Agricultural Research
      The Organization of Science for Maximum
         Effectiveness

BIOLOGICAL NITROGEN FIXATION: A FERTILIZER STRATEGY
POTENTIALLY BENEFICIAL FOR THE POOR IN DEVELOPING
COUNTRIES -- Marvin R. Lamborg 115

Nitrogen Fixation by Pure Cultures of
Rhizobium
Glutamine Synthetase in Rhizobium
Experimentally-Induced Bacteroids?
Organic Nitrogenous Compounds Synthesized
for Transport and Assimilation
The Recognition/Infection Process in Legumes
Relationship Between Photosynthesis and
Nitrogen Fixation
Efficiency of Nitrogen Fixation in Legumes
Characteristics of the Molybedenum Iron
Cofactor of Nitrogenase
Non-legume, Symbiotic Nitrogen-fixing
Systems: Alnus
Non-legume, Symbiotic Nitrogen-fixing
Systems: Azolla
Concluding Remarks

TRANSLATING BASIC RESEARCH ON BIOLOGICAL NITROGEN
FIXATION TO IMPROVED CROP PRODUCTION IN LESS-
DEVELOPED COUNTRIES -- A USER'S VIEW -- R.W.F.
Hardy 137

Introduction
General Approaches to Meeting Nitrogen Needs
Advances and Impacts of Nitrogen Input Research
The What's Wrong With and Opportunities for
Improvement of Biological N_2 Fixation
Steps in Legume N_2 Fixation
Possible Future Technologies
Concluding Thoughts

LINKING BASIC RESEARCH TO CROP IMPROVEMENT PROGRAMS FOR
THE LESS-DEVELOPED COUNTRIES: BIOLOGICAL CONTROL OF
INSECTS -- Thomas R. Odhiambo 153

The Scale of the Rural Problem
Mixed Cropping Entomology
Conventional Attitude to Crop Resistance
Insects as Chemists
The Future of Crop Protection in the Developing Countries

USE OF PREDATORS AND PARASITOIDS IN BIOLOGICAL CONTROL --
  C.B. Huffaker   173

      Introduction
      The Empirical Basis of Biological Control
      The Theoretical Basis of Biological Control
      Some Classical Examples Illustrating Ecological
        Principles
      Manipulations to Conserve and Augment Resident
        Natural Enemies
      Programs Integrating Chemical and Biological
        Control

TOWARD MORE RATIONAL POLICY -- D. Woods Thomas   199

      Introduction
      Technical Constraints to Increased Resource
        Productivity
      The Role of Research
      The Dilemma
      Why the Dilemma?
      The Bottom Line

THE ROLE OF SCIENTIFIC RESEARCH IN RURAL DEVELOPMENT --
  Milton J. Esman   209

Contributors   217

Author Index   219

Subject Index   225

LINKING BASIC RESEARCH TO CROP IMPROVEMENT PROGRAMS IN DEVELOPING COUNTRIES

Ruben L. Villareal

Asian Vegetable Research and Development Center
Tainan, Taiwan, ROC

and

Visiting Professor
Department of Plant Breeding and Biometry
Cornell University
Ithaca, New York 14853

I was born on the Asian continent, a continent inhabited by more than 2 billion people. Its population growth rate is high and food supply is scarce. Consequently, malnutrition and starvation are widespread and rampant and there is no likelihood that food problems will be solved in the very near future. That is why as an insider and scientist from the Asian continent I can speak of how and what basic studies conducted in the affluent nations can be more useful to crop improvement programs in developing countries. For the same reason I know also what other things are of greater value to us.

AGRICULTURAL RESEARCH IN DEVELOPING COUNTRIES

But first of all let me draw a clearer picture of agricultural research in developing countries. In my opinion both the number of people supplied by a farm worker and the yield per hectare are good measures of what agricultural research has done for a country.

Developing countries lag behind the developed countries in the number of people that can be supplied by a farm worker (Table 1). In China, for example, a farmer in 1977 can supply the needs for agricultural products for himself and of 1.61 other persons.

Table 1. Number of people supplied by a farmer in
          selected developed and developing countries,
          1965 compared with 1977.

Country	1965	1977	Change (%)
Developed			
Japan	3.79	7.58	100
UK	29.41	45.45	54
USA	19.60	40.00	104
Developing			
China	1.40	1.61	15
Egypt	1.77	1.93	9
Mexico	1.99	2.58	30

Source: 1977 FAO Production Yearbook.

Whereas in the same year an American farmer with the help of those
who provide him with information, seeds, fertilizers, and other
production inputs can supply food and clothing for himself and 40
others. This leaves more people to do other jobs (i.e. teacher,
factory worker, carpenter, street cleaner, doctor, lawyer, or
engineer). For the same period of time, the rate of change in how
many people a farmer can serve in Japan and the United Kingdom was
54 and 100 percent, respectively. It was only 15 percent in China,
9 percent in Egypt and 30 percent in Mexico. Crop productivity in
developing countries was much lower than the developed countries
(Table 2). Many factors contribute to this condition. The most
important factor, however, is lack of support for agricultural
research. Cummings (1976), for example, stated that the national
investment in support of agricultural research, education, and
production services in developing countries is low by any measure
(i.e. percentage of GNP, per capita per year, or value of agricultural production). For example, in 1965 the research expenditures
as percent of value of agricultural production were 0.87 percent in
developed and 0.26 percent in developing countries. The comparative
growth of expenditures on agricultural research in millions of U.S.
dollars show that in 1951 developed countries spent about 8.3 times
more than developing countries. And in 1970 the gap of expenditures
became even wider (Figure 1). These countries suffer from lack of
trained staff who will conduct research and extend information to
farmers. The few trained ones in most cases don't have the incentives because of inadequate salaries and support. Therefore, most
of these countries need external assistance to partially meet the
needs for trained manpower and badly needed technology for agricultural production.

Table 2. Comparative crop productivity of developed
          and developing countries, 1977.

Crop	Developed (kg/ha)	Developing	Difference
Cereals			
Rice	5476	2467	3009
Corn	4724	1655	3069
Wheat	1924	1298	626
Vegetables			
Cabbage	25726	13144	12582
Potato	15717	10397	5320
Tomato	27518	14566	12952
Legumes			
Soybean	1946	1186	760
Drybean	634	500	134
Peanut	2351	872	1479

Source: 1977 FAO Production Yearbook.

As expected when resources are limited, research emphasis in developing countries has been on cereals. For this reason, progress in improving productivity is faster in cereals than in vegetables. Average yearly increase in the yield of cereals ranged from 1.7 to 3.2 percent whereas that of vegetables was from 0.4 to 1.3 percent (Villareal, 1978). In addition, research by the International Rice Research Institute (IRRI) and the International Maize and Wheat Improvement Center (CIMMYT) established in 1960 and 1966 respectively, contributed heavily to the increased productivity experienced. The Asian Vegetable Research and Development Center, on the other hand, was established only in 1971. The relatively low priority on other crops such as vegetables is understandable since most of these countries need staple food to meet the demands of their rapidly increasing population. In many instances, a government that can assure abundant supply of cereals such as rice and corn is likely to stay in power longer.

Typically, applied research is the major emphasis in developing countries, since what is needed are those studies that will pay off in a shorter time. Most studies therefore are problem-oriented. Where there are strong colleges of agriculture and research institutes, however, both applied and basic studies are usually carried out by graduate students and qualified staff, respectively. In many developing countries, private industries with the backing of the government establish research institutes for specific commodities with strong export market. For instance, commodity institutes such as the Malaysian Rubber Institute, the Philippine

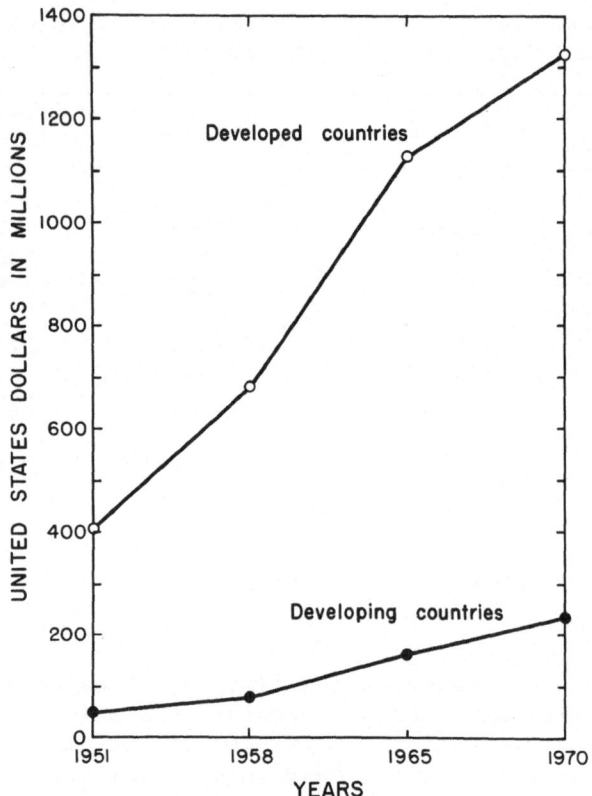

Figure 1.   Comparative expenditures on agricultural research
            between developed and developing countries. (Source:
            R.W. Cummings, Jr. 1976)

Sugar Research Institute, and the Banana Research Institute (Taiwan)
have been established and conduct both applied and basic research.

APPLICATION OF BASIC RESEARCH IN CROP IMPROVEMENT

     Basic studies in genetics, cytology, physiology and bio-
chemistry have contributed a wealth of information that led to
numerous breakthroughs in crop improvement. Some relevant studies
will be mentioned in this paper.

     The self-pruning (sp) character in tomato appeared as a spon-
taneous mutation in Florida in 1914 (Rick, 1978). The gene has
been used extensively in tomato improvement programs throughout
the world. One of the first successful uses of this mutant was

its transfer to Sta. Clara Canner, an indeterminate processing
variety. Determinate progenies from this cross resulted in Pearson
varieties which became the predominant canning varieties in California from the early 1940's to the 1960's (personal communication
with H.M. Munger, Cornell University). The same gene was also
essential to further variety improvements that revolutionized and
saved the tomato industry in California when labor became too
scarce for hand-picking. The gene sp causes the plant to grow in
an orderly, compact, determinate fashion, flower more abundantly,
and ripen in a shorter season than the normal (indeterminate),
thus allowing machine harvesting of tomatoes. In 1977, California
harvested 80 percent of the 7 3/4 million tons of processing
tomatoes produced in the U.S. (Stevens, 1978) -- a feat impossible
to accomplish without machine harvesting as labor in the U.S. is
scarce and expensive. The same gene permits Asian farmers to
intercrop tomato with sugarcane and fruit trees.

Equally dramatic is the transfer of disease resistance genes
from wild Lycopersicon species to acceptable improved cultivars
(Walter, 1955). Without such genes, growing of tomatoes anywhere
in the world today would not be as profitable.

The dwarf characters in rice and wheat which are both controlled by a single recessive gene facilitated rapid breeding programs in these crops and changed plant architecture which resulted
in the development of several high yielding varieties that are grown
now in more than 20 million hectares (Chandler, 1973). The dwarf
traits played a significant role in ushering the "Green Revolution"
and feed most of mankind.

As early as 1962, Filipino scientists conducted a basic study
to demonstrate the cytoplasmic inheritance of susceptibility to
Helminthosporium maydis (Villareal and Lantican, 1965). They
predicted that the corn industry in the U.S. which used the T-
cytoplasm would suffer great loss if environment would be favorable
for the development of the pathogen. The economic significance of
such a study, however, was not realized until the disease destroyed
a large part of the U.S. corn crop in 1970.

The availability of cytoplasmically inherited susceptibility
to H. maydis permits hundreds of basic studies in mitochondrial
biochemistry and ultrastructures. Dr. P. Gregory and his col-
laborators (1978) at Cornell Univeristy showed that the critical
interaction between host and parasite in southern corn leaf blight
disease involves a reaction between HmT toxin and T mitochondria.
Their observation of the only mitochondrial mutant in higher plants
have led them to initiate a series of fundamental studies on higher
plant mitochondrial genetics. This subject will lead to better
understanding of cytoplasmic genes which is of tremendous scientific
and economic importance.

Physiologists at the Asian Vegetable Research and Development Center found that Chinese cabbage seeds treated with 5 ppm of gibberellin A 4/7 for 24 hours before placing them on agar medium for germination provides a selection criterion to identify heat-tolerant varieties, which have significantly longer hypocotyls than heat-sensitive varieties following treatment (AVRDC, 1976). The finding has become a rapid and useful technique to screen for heat tolerance in Chinese cabbage during the seedling stage.

Another interesting fundamental study in the field of bio-chemistry, physiology and genetics is that of self-incompatibility in higher plants. Ferrari and Wallace (1976, 1977 a&b) at Cornell have published a series of excellent papers on the subject of better understanding of self incompatibility in the Brassicas. Results of their studies provide an efficient and effective procedure of identifying highly self-incompatible S-allele homozygous lines. Hopefully, the cost of seed production of hybrid cabbage and broccoli, which is totally dependent upon use of self-incompatibility as the mechanism for ensuring cross fertilization between inbred lines, will become cheaper as a consequence of their studies. Eventually, the cost of hybrid seeds will be within the reach of small farmers in developing countries. In addition, better under-standing of self-incompatibility could lead to its utilization in other economically important plant species.

Biochemists have found the opaque-2 mutant gene in corn which approximately doubles the amount of two limiting amino acids lysine and tryptophan. Thus, the quality of maize protein may be improved by transferring this gene to a normal corn. Unfortunately the mutant carries with it some undesirable genes. CIMMYT scientists believe that a breakthrough in commercial use of high quality protein maize will come when a variety with opaque-2 gene shows performance equalling or surpassing the existing normal varieties (CIMMYT, 1975). If accomplished, corn-eating people will be greatly benefited as will the animal feed industry.

AREAS OF GREATER VALUE TO DEVELOPING COUNTRIES

For developing countries to best utilize results from funda-mental studies such as in photosynthesis and cell biology, they must first have found an application in developed countries or International Agricultural Research Centers (IARC). Even after this stage is attained, their eventual application and usefulness will vary from country to country depending on the economic growth and availability of trained manpower who will do the job. In essence a second stage technology application of basic research will find its adoption in developing countries faster. For example, let us assume that a mutant gene that accelerates photosynthesis

or non-legume plant that can fix nitrogen is found from a wild crop
species. Scientists from either IARC or developed countries can
transfer the gene from the wild parent into a genetic background
of the cultivated species. Thus they provide material that is far
more useful to their fellow scientists in developing countries
than the wild accession in which the mutant gene was first observed.
Of course, adequately trained scientists from developing countries
can do the same but their limited resources would rather be spent
in achieving immediate goals with the least expenditure of time and
effort.

It should be pointed out, however, that a direct transfer of
technology from developed to developing countries is also possible
in some exceptional cases. For example, in 1969 Dr. H.M. Munger
of Cornell University introduced into the Philippines a muskmelon
variety called Gulfstream which is resistant to downy mildew. At
that time this disease was inflicting considerable damage on the
melon industry of Candaba swamp. An interim recommendation was
made by the University of the Philippines to use this variety which
was readily accepted by the growers. There was a time that 100
percent of the melon area in Candaba swamp was planted to Gulfstream.
This example illustrates that even without an intensive screening
for disease resistance and yield, an outstanding variety may be
recommended in a developing country for commercial planting
especially if that variety is markedly superior to the variety
grown by the farmers. Another outstanding example is a hybrid
cabbage from Japan which found large-sized heads in an experimental
plot of the university at the time when the Philippines recorded
the highest summer temperature. KK, the hybrid cabbage, was
recommended for summer planting in the lowlands. Today 100 percent
of lowland cabbage in the Philippines is grown to KK. That same
F_1 hybrid has been adopted in many lowland production areas of the
tropical zone (Villareal et al., 1972).

Perhaps of greater value to developing countries as far as
basic studies are concerned are those studies that will result in
the development of crop species with more stable resistance to pests
and diseases, tolerance to excessive water and drought, tolerance
to adverse environmental stresses, efficient uses of fertilizer
and nutritious quality. Of equal importance are basic studies
that will support the development of management practices appropriate
for whatever crop varieties will be bred.

Nevertheless even with improved varieties and appropriate
management technology, seed production and distribution systems must
be either improved or developed. This is one field which has always
been neglected but could be an excellent vehicle to bring into use
the results of crop improvement. Any efforts of crop improvement
and agricultural production in general will be useless until the

rural people themselves participate in the action as producers.
To make agricultural production efforts a reality, local governments
need to introduce some practical innovations such as credit and
cooperatives, irrigation systems, roads from farms to markets, means
of communication and marketing systems for agricultural produce
including processing and storage facilities.

Finally, results from basic research will surely contribute to
science and practice in both developed and developing countries.
The former can better try them first and when their application is
a certainty they can be turned over to the latter for experimentation
and adoption. Today some developing countries are probably ready
to conduct more basic research if funds are not limited. Most of
them, however, need more practical innovations to make their rural
poor produce food and other agricultural products.

REFERENCES

Asian Vegetable Research and Developmenter Center. 1976. Annual
    Research Highlights '75. Shanhua, Taiwan, ROC. 23p.
Centro International de Mejoramiento de Maize y Trigo. 1975. CIMMYT
    review 1975. El Batan, Mexico.
Chandler, R.F. 1973. The scientific basis for the increased yield
    capacity of rice and wheat, and its present and potential impact
    on food production in the developing countries. In: Poleman,
    T.T. and D.K. Freebairn. 1973. eds. Food, Population and
    Employment - The impact of the green revolution. 272 pp.
    Praeger Publishers, New York.
Cummings, R.W., Jr. 1976. Food crops in the low-income countries:
    The state of present and expected agricultural research and
    technology. Working papers of The Rockefeller Foundation.
FAO. 1978. 1977 FAO Production Yearbook. Rome.
Ferrari, T.E. and D.H. Wallace. 1976. Pollen protein synthesis and
    control of incompatibility in Brassica. Theor. Appl. Genet.
    48: 243-249.
Ferrari, T.E. and D.H. Wallace. 1977a. Incompatibility on Brassica
    stigmas is overcome by treating pollen with cycloheximide.
    Science 196: 436-438.
Ferrari, T.E. and D.H. Wallace. 1977b. A model for self-recognition
    anf regulation of the incompatibility response of pollen. Theor.
    Appl. Genet. 50: 211-225.
Gregory, P., D.E. Matthews, D.W. York, E.D. Earle and V.E. Gracen.
    1978. Southern corn leaf blight disease: Studies on mito-
    chondrial biochemistry and ultrastructures. Mycopathologia
    66: 105-112.
Rick, C.M. 1978. The Tomato. Scientific American 239: 76-87.
Stevens, M.A. 1978. Breeding tomatoes for processing. Paper pre-
    sented at the First International Symposium on Tropical Tomato
    held at AVRDC, Taiwan, ROC on Oct. 23-28, 1978.

Villareal, R.L. and R.M. Lantican. 1965. Cytoplasmic inheritance of susceptibility to Helminthosporium leaf spot in corn. Phil. Agric. 49: 294–300.

Villareal, R.L., V.G. Balaoing, D.T. Eligio, L.Z. Magate, I.C. Catedral, F.P. Agbigay, and W.A. Dancel. 1972. Choosing varieties and planting dates. In the Philippines Recommends for Vegetables. 1972–1973. p.6–9.

Villareal, R.L. 1978. Tomato production in the tropics – Problems and progress. Paper presented at the First International Symposium on Tropical Tomato held at AVRDC. Taiwan, ROC on Oct. 23–28, 1978.

Walter, J.M. 1955. Hereditary resistance to disease in tomato. Ann. Review of Phytopathology 5: 131–162.

THE EVALUATION AND REMOVAL OF CONSTRAINTS TO CROP PRODUCTION

N. C. Brady
Director General
International Rice Research Institute (IRRI)
Los Baños, Laguna, Philippines

My orientation and that of the Institute I represent is toward food production and how research can be directed to help increase that production. Events of the past two decades suggest that whether we like it or not, we all must be similarly orientated. The current state of relative world food adequacy can be likened to the Biblical years of plenty, certainly to be followed by similar years of food shortage and famine.

The opinion that the world's food problems will continue to be critical is shared by competent analysts and review panels (IFPRI, 1977; ADB, 1977; Brown, 1978; Eckholm, 1976). For example, economists at the International Food Policy Research Institute predict that by the year 1990 production of staple food crops in developing countries will fall short of food demands by from 120 to 145 million metric tons (t), 3 times the 37 million t shortfall of 1975 (IFPRI, 1977). Hardest hit will be the poorest countries with per capita gross national product (GNP) in 1973 of less than $300. Their food deficit will be 75 to 80 million t/year.

An Asian Development Bank (ADB) study team that compared the 1977 food situation in South and Southeast Asia with that of a decade earlier stated "Overall, the most optimistic view which can be taken of the food situation is that the region is not much worse off now than at the time of the first Asian Agriculture Survey" (ADB, 1977).

The sobering effects of steadily and rapidly increasing rates of population growth rates on food sufficiency can be seen from data on the change in grain trade in Mexico over the past two decades (see Figure 1). In the early 1960s, Mexico was heralded as a net exporter of grain, a happy consequence of the green revolution spawned largely

by the new wheats developed in that country. A decade later, the
temporary respite disappeared, and Mexico is now a major grain
importer. Production increases have not equalled the rapidly rising
population in Mexico.

The enormity of the task ahead is illustrated by the fact that
if a reasonable 2.5% annual increase is to be attained over the next
25 years, annual cereal production must increase from the current 400
million t level to 950-1000 t (Aziz, 1977). To reach this goal, the
rate of increase in food production must be near that which characterized the "green revolution" of the past 15 years. A similar level
of research must be called on to provide the technology bases for this
increased food production.

UNIQUE SIGNIFICANCE OF ASIA AND OF RICE

No region of the world is entirely free from the problems of
providing adequate food for its population. But because of sheer
numbers of people to be fed, the magnitude of Asian food problems
dwarfs that of other continents. An estimated three-fourths of the
people who receive insufficient calories or proteins, or both, reside
in Asia (Table 1). Furthermore, population growth in Asia, although
it shows signs of slackening somewhat in some countries, is high
enough to add 40 to 45 million new mouths to be fed each year. Food
for this growing population must be produced on land now under cultivation because most tillable areas are already being farmed.

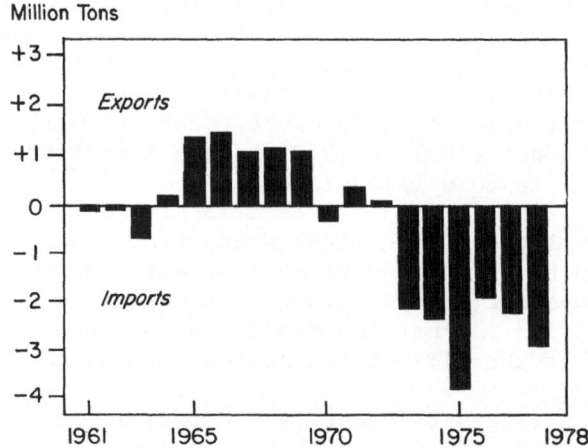

Fig. 1. Mexico: Net Grain Trade, 1961-78.
Sources: Brown, 1978; USDA, 1974.

Table 1. Estimated magnitude of world malnutrition, 1970.

| Region | Population (billions) | Population Affected by Insufficient Protein/Energy Supply | |
		Percent	Number (millions)
Developed Areas	1.07	3	28
Developing Areas*	1.75	25	434
Latin America	0.28	13	36
Far East	1.02	30	301
Near East	0.17	18	30
Africa	0.28	25	67
World Total*	2.83	16	462

*Excluding Asian centrally planned economies. Source: Knight and Wilcox, 1976 from USDA, 1974.

RICE AS THE FOOD STAPLE OF THE POOR

Rice is the primary food staple of most Asians and of low-income people in most tropical and semitropical countries. Furthermore, rice will probably continue to be their primary caloric source, particularly in the humid tropics where it is the only crop which will produce successfully in the monsoon season.

Fortunately, rice is perhaps the world's most adaptable food crop, grown under more variable environmental conditions than any other major crop. Rice produces well under a wide range of environmental conditions, varying from that of an upland crop to wetland areas with flooded water levels of up to 5 meters in depth. It is grown as far north as 53° latitude in China and 35 to 40° south in Australia. Its tolerance for temperature extremes permits production in the irrigated desert areas of Pakistan and Iran with mean temperatures of 35°C and in the hill areas of Nepal with mean temperatures of 15 to 18°C. Rice tolerates a wide range of soil pH conditions. It is grown on the acid sulfate soils of Thailand, Vietnam, and Burma, and on the alkaline and saline soils of India and Pakistan. Rice is a remarkable crop.

Because of its unique role as a food for 90% of the world's truly poor and its wide adaptability to different ecologic zones, rice will receive primary attention in the sections that follow.

THREE SOURCES OF MORE FOOD

There are three obvious ways to produce more food:
1. to bring more land under cultivation,
2. to increase crop yields on the land now cultivated, or
3. to increase the number of crops produced each year on old and new lands.

Producing crops on previously untilled lands has been the primary means of increasing world food supplies for centuries and still offers considerable potential for expansion in Africa and Latin America. But in Asia, where food production increases in the coming decades must be largest, most tillable lands are already farmed. In fact, increasing demands for the use of farm lands to accommodate steadily increasing populations such as for communities, roads, and industrial sites may actually reduce the productive cultivated area in some countries.

The potential to increase crop yields on land now being cultivated is great, especially in the tropics. Table 2 shows the difference between the potential rice yields and those that farmers in the temperate and humid tropical regions now actually produce. If actual rice yields in the tropics could be raised to 20 or 25% of the potential yields (as has been done in temperate regions), there would be no rice shortage for some time. We will deal further with this matter later on.

The third means of increasing crop production is by increasing cropping intensity -- merely growing more crops per year per land area. Once again the humid tropics have great potential for improvement -- there is no winter and rainfall is high at least during part of the year. Bradfield (1970) pointed out that the cumulative growing degree days (using a base of 50°F) for Los Baños, Philippines, is over 11,000, more than 4 times the comparable value for Ithaca, New York (2500). He demonstrated in experiments that four or five crops could be grown yearly at Los Baños. Since then, practical farm cropping systems have been developed to permit even small landholders to produce four crops of rice per year, or two of rice and one or more of upland crops (IRRI,1978). Using the four-crop rice system, IRRI scientists have produced 23.3 t/ha during 1 year. This system provides the farmer with remarkable increases in net return (Table 3).

Obviously a number of factors, including research findings, will determine the degree to which each of these three mechanisms will

Table 2. A comparison among potential and actual rice yields (tons per hectare) in countries in temperate and humid tropical climates.

	Temperate Zone	Humid Tropics
Potential crop yields	15–17	13–15
Actual crop yields	4.5–6.0	1.5–2.5
% Actual of Potential	25–40	10–20

Table 3.   Comparison of the costs and returns of a farmer's double
           cropped rice and of a continuous rice production pattern
           which produced nearly 4 rice crops a year.

Cropping Pattern	Variable Costs	Total Returns	Net Returns
Farmer's double rice crop	404	650	246
Continuous rice production (3.96 crops/yr)	1618	3162	1544

Source: IRRI, 1978.

increase future food production.   IRRI scientists have estimated the
probable future increase in rice production they expect research
findings will make possible.   Table 4 shows anticipated increases due
to higher yields, cropping intensities, and irrigation under different
rice cultural conditions.   The researchers expect about 55% of the
future rice production increase from increased yields and 45% from
increased cropping intensities.   They also expect almost half the
increases in production to be on irrigated land and another fourth
on rainfed shallow areas.   Those estimates should be useful in setting
research priorities to remove production constraints.

THE POTENTIAL-PERFORMING GAP

    In some tropical and semitropical areas, progress in increasing
crop production during the past decade has been remarkable.   For
example, wheat yields increased dramatically in India and Pakistan
as did rice yields in Colombia (Jennings, 1976).   But in most tropical
countries, rice yields increased only modestly even after the fertilizer-responsive semidwarf varieties were released.   Furthermore,
farmers have adopted the modern rices in only 25 to 30% of Asia's
rice-growing areas.   Adoption in Burma, Thailand, and Bangladesh is
only about 7, 11, and 14% respectively (Dalrymple, 1978).   Obviously,
the green revolution has had spotty success -- good in some regions
but failing in others.

    Data from two sections of India help explain the slow rate of
increase of rice yields in much of Asia (Figure 2).   In the highly
productive northwestern area, yields have increased dramatically
since the coming of modern rice varieties.   In fact, the rate of
increase actually exceeds that for wheat during a comparable time

Table 4. The percentages of future benefits from research that can
be expected from yield increases and from crop intensification under different environmental complexes.

Environmental complex	Future research benefits (%) expected from		
	Yield increases	Crop intensification	Total
Irrigated	22	24	46
Rainfed (shallow)	15	11	26
Rainfed (medium deep)	6	2	8
Deep water	2	1	3
Upland	3	1	4
Arid, high temperature	6	5	11
Long day, low temperature	1	<1	1
	55	45	100

Source: IRRI, 1979.

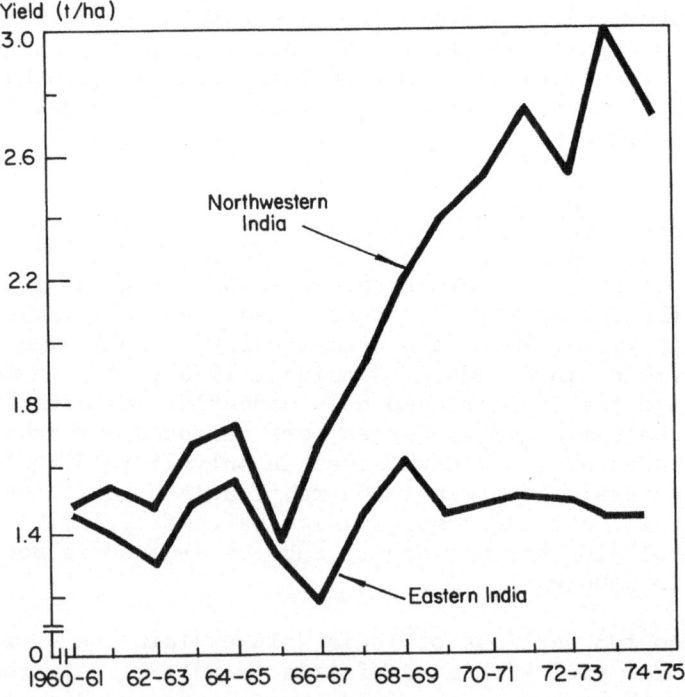

Fig. 2. Rice yield trends for eastern and northwestern India since
1961. Source: IRRI, 1978.

period. In contrast, average yields in eastern India remained
essentially unchanged from 1960 to 1975. Yield levels in these two
regions are associated with differences in adoption of improved varieties and of the technology that permits such varieties to perform.
The new varieties were adopted in the northwest where irrigation is
assured and disease and insect pressures are low. But they were not
adopted in the east where flood control is poor and pest pressures
are high. So far, no modern varieties have been developed that will
produce well in the flood-prone areas of eastern India and little
research has been conducted to develop such varieties.

CONSTRAINTS ON RICE YIELDS IN THE TROPICS

Many factors account for the marked differences in rice yield
levels, first between the rice yields of temperate and tropical countries, and second, among yields of different regions within the
tropics. Socioeconomic factors account for some of the differences.
Price policies in Japan, Korea, and Taiwan, for example, encourage
high inputs of fertilizer and pesticides. Such incentives are less
common in most tropical countries (Table 5). Obviously, the farmer
in Korea or Taiwan (Republic of China) has much more economic incentive to use fertilizer than the Thai farmer has.

Most rice-growing areas in the temperate countries also have
dependable irrigation water through government projects. But most
tropical countries lack such dependable water sources. Such socioeconomic constraints, although important, are beyond the major thrust
of this paper. They are subject to research, however, and their
removal will be essential if the benefits of physical and biological
research are to be realized.

Table 5. Ratio of the price of urea nitrogen compared to the price
of paddy rice for selected countries in Asia.

Country	1970	1972	1974	1976
Burma	5.33	–	–	1.81
China, Republic of	2.23	1.80	–	0.78
Korea	1.07	0.74	0.77	1.42
Philippines	3.81	2.33	4.07	3.56
Sri Lanka	1.33	1.27	–	2.04
Thailand	8.10	9.44	9.16	4.08

Source: IRRI, 1976.

The following physical and biological stresses that constrain
rice yields in the tropics are most significant:

1. Water supply
   a) Excesses (floods)
   b) Deficiencies (drought)
2. Pest damage
   a) Insects
   b) Diseases
   c) Weeds
3. Adverse soils
   a) Toxicities
   b) Deficiencies
4. Temperature extremes
   a) Low temperatures
   b) High temperatures

At first glance, one might suggest that this list is no different for the tropics than for the temperate zones, and that is true.
But the stresses tend to be more severe in the tropics, for a number
of reasons. Specific examples will be given for each type of stress.

Water Supply - Floods

Floods are common in tropical areas with monsoon climates. In
some areas, including eastern India and parts of Bangladesh, Burma,
Thailand, and Vietnam, the annual floodwaters cover millions of hectares at depths of from 0.3 to 2 or even 5 meters. For most of these
regions, there are neither high-yielding varieties nor associated
modern technologies. Even in areas where the annual floodwaters reach
levels of only 0.5 to 1 meter, the short-statured rice varieties tailored for controlled water conditions succumb when submerged and are
inferior to traditional varieties. Likewise, such common farmer practices as topdressing the crop with nitrogen fertilizer are fruitless
in standing water of 0.5 meters or deeper. No wonder farmers in these
areas have not adopted the "modern technologies."

An estimated 25% of the rice areas of South and Southeast Asia
is subject to annual floods sufficiently deep to prevent the use of
the current range of high-yielding varieties (IRRI, 1979). In even
wider areas floods occur often enough to make it risky to plant varieties that cannot tolerate temporary submergence. It is easy to see
why farmers of eastern India stick to their traditional varieties.
Science has given them nothing better suited for their circumstances.

One social constraint facing farmers in flood-prone areas is
the fact that few experiment stations are located in or near those
areas. Furthermore, the probability of success of research to produce
new varieties and technologies for flood-prone regions tends to be

lower than for areas with good water control. Consequently, little
research has been conducted to develop better varieties and technologies for flooded areas, and the researchers assigned to such tasks
are generally less well-trained and motivated. That illustrates the
interrelationship between the "people" problems and the "plant production" problems.

Water Supply - Deficiencies and Drought

It seems illogical or contradictory to rate drought as a major
constraint on rice yields in the humid tropics. But the "humid"
adjective refers to total rainfall, not to its distribution. A series
of high intensity rainfalls total up to high yearly precipitation
levels and cause excessive surface runoff and percolation. Between
the downpours, however, are rainless days or weeks when high temperatures and high solar radiation bring drought stress.

Drought is especially serious in upland rice areas but in most
years drought also adversely affects rice production in rainfed and
in partially irrigated areas. Even the flood-prone areas where floating rices will prevail later in the year, are subject to drought imme-
diately after seeding. Consequently, moisture stress may limit yields
at some time during the growing season on all but the 15 to 25% of the
rice lands with dependable irrigation.

The frequency of drought damage in Thailand from 1907 to 1965 is
shown in Table 6. Note that drought was the primary cause of crop
damage in 6 of Thailand's 9 disaster years. These data illustrate
the importance of drought in limiting crop production even in the
humid tropics.

Damage from Insects and Diseases

Disease and insect pests reduce rice yields drastically in the
humid tropics -- particularly in South and Southeast Asia where rice
has been grown continuously for centuries. The area has no cold
winter months to reduce pest populations. The rice monoculture, which
in the past has been dictated by the monsoon climate and inadequate
technology, may encourage pest buildups. The introduction of high-
yielding varieties has probably increased the pest pressures. The
dense canopy encouraged by high-tillering and fertilizer-responsive
varieties provides an environment that encourages some diseases and
insects.

Scientists from IRRI and from cooperating national programs have
developed varieties with resistance to six or seven of the most serious
pests. But even where these varieties are used, damage from minor
pests can seriously limit rice production and the introduction of a

new disease or insect can wreak severe damage.

Adverse Soils

     As researchers move out from their experiment station plots onto
farmers' fields the yield constraints posed by adverse soil conditions
become more apparent. Experiment stations are generally located on
good soils. Even if the experiment station soils are poor or toxic,
the deficiencies or toxicities on both can easily be alleviated by
using public funds. But the low-income farmer is stuck with what he
has, and most are located on soils that are inferior to those on
experiment stations or government farms.

     In recent years there has been a growing recognition of the
degree to which soil deficiencies and toxicities limit crop production
in the tropics. The need to meet deficiencies of the major mineral
elements, especially nitrogen and phosphorus, has long been recognized. Even so, the quantity of chemical nutrients being applied per
hectare in most developing countries is low (Table 7). In the LDCs
of South and Southeast Asia, rates of N, P, and K use are generally
less than 30 kg/hectare per year. Even in countries such as Pakistan
and the Philippines, where the adoption of modern varieties is high,
fertilizer application rates are a fraction of those that research
suggests are economical.

Table 6. Damaged area, production, and cause of damage to rice in
          Thailand (1907-1965).

Year	Damaged area (%)	Production (thousand t)	Production decrease from previous year (%)	Cause of damage
1917	21.0	2989	21.1	Flood
1919	43.4	2270	32.9	Drought
1928	16.3	3882	14.1	Drought
1929	19.5	3875	0.2	Drought
1936	31.7	3380	28.5	Drought
1942	34.3	3854	24.7	Flood
1945	24.3	3572	27.5	World war
1954	18.6	5709	30.7	Drought
1957	15.5	5570	32.9	Drought

Source: Isrankura, 1966.

Table 7. Total fertilizer consumption per hectare for selected LDCs.

Country	Fertilizer Consumption (kg of NPK/ha)	
	1963[a]	1973[b]
South Asia		
Bangladesh	4.7	21.2
India	3.7	16.5
Nepal	0.4	6.8
Pakistan	3.4	22.2
Sri Lanka	11.2	51.6
Southeast Asia		
Burma	1.0	6.7
Cambodia	0.4	1.1
Indonesia	7.3	21.1
Laos	0.1	0.2
Malaysia	14.8	–
Philippines	11.0	22.1
Thailand	2.5	12.0
Vietnam	17.6	52.6
East Asia		
China, Republic of	225.8	340.8
Korea	157.9	319.1

[a]Average of 1961–65. [b]Average of 1972–74. NPK = Nitrogen, Phosphate, and Potash. Source: FAO, 1976.

As NPK chemical fertilizers are more widely used, a lack of
response to the application of these major elements calls attention
to the prevalence of deficiencies of other essential elements in the
tropics. Zinc deficiency is "perhaps the most common deficiency
after nitrogen" in wetland areas (Ponnamperuma, 1977). Sulfur deficiency is also increasing (Blair et al., 1978), especially in areas
with increasing NPK usage, with low soil sulfur, and with low additions
of sulfur from industrial air pollution.

Soil toxicities also provide serious constraints to crop production throughout the tropics (Sanchez and Buol, 1975; Spain et al.,
1975). Salinity is perhaps the most common of these mineral toxicities in Asia, although excess alkalinity or acidity, iron toxicity,
aluminum toxicity, and low productivity of peat soils are also important. Adverse soils have been estimated to limit the production of
both native and modern rice varieties on some 40 million hectares and
a similar area, climatically and physiographically suited to rice,
lies uncultivated because of adverse soils (IRRI, 1975). Examples of
the latter are the large areas of acid sulfate soils in Vietnam and

Thailand and similarly vast regions of saline soils and tidal swamp areas in several Asian countries.

In seeking solutions to the constraints that adverse soils place on crop production, the economics of soil improvement must be kept in mind. Farmers with meager financial resources cannot afford high resource inputs to improve their soils. Consequently, initial research efforts are being placed on the development of varieties and technologies that will tolerate and yield well under the adverse soil conditions. This will receive attention in later sections of this paper.

Temperature Extremes

The temperatures of most tropical areas are favorable for rice production throughout the year (25–35°C). But at high elevations, low air and water temperatures limit crop yields. These low temperatures may also extend the growth duration excessively, thereby reducing the potential to use some of the high-yielding varieties developed in the lowlands. Temperatures lower than 20°C adversely affect most phases of rice growth (Chang and Oka, 1976).

In some irrigated desert areas, temperatures are high enough to induce sterility especially during midday when anthesis of most varieties occurs. Maximum day temperatures higher than 40°C are common in Pakistan, Iran, and the Middle East during periods of the year when other factors favor rice production. Since daytime temperatures of higher than 35°C at flowering may increase spikelet sterility (Yoshida, 1978), the yields of most varieties are reduced at these high temperatures. Varieties are being sought that will tolerate high summer temperatures.

RESEARCH TO REMOVE CONSTRAINTS

Research to remove constraints resulting from plant stress can be divided broadly into two groups: (a) those aimed directly at eliminating the stress, and (b) those aimed at changing or improving the rice plant to accommodate the stress. The priority approach in a given situation is determined by the feasibility of solving the problem and by the probability that low-income farmers are able to apply those findings. The meager resources of farmers in the tropics is a primary factor in making this decision. Generally speaking, it is more feasible to change the plant than to try to remove the stress.

RESEARCH TO ELIMINATE THE STRESS

The best example of research efforts to remove the stresses are those concerned with water and plant nutrients. Water management research focuses on the plant and its response to water, but also takes into account soil management factors and weaknesses in irrigation systems. Once again, the interaction of biophysical and socioeconomic factors is evident. Efficiency of irrigation water utilization, even in areas of low water supply, is woefully low in the tropics, largely because of "human" factors. The water is there but its distribution and, in turn, effective use by the rice plant is faulty. Research is under way to determine what can be done to help low-income farmers increase the efficiency of irrigation water use.

Through cropping systems research, the potential to increase the efficiency of water utilization through soil and crop management becomes more apparent. For example, in regions with distinct wet and dry periods, marked savings in soil moisture have been noted through the establishment of a "soil mulch" at the end of the wet season (IRRI, 1978). The soil mulch concept, much in vogue half a century ago but later largely discarded as having little applicability in temperate climates, can now be reexamined. Moisture saved by a soil mulch will permit the early establishment of rice or other crops -- an important consideration in crop intensification schemes.

Japanese and Indian researchers have shown that root zone placement of nitrogen fertilizers doubles the efficiency of the rice plant's nitrogen use. Recently researchers in six tropical countries have confirmed these findings (IRRI, 1978a). This is of great significance to a farmer whose annual income may be $500 or less. Likewise, practical methods are being sought to utilize blue-green algae or the aquatic fern Azolla to provide biologically fixed nitrogen for this farmer. Such nitrogen sources could be even more important if prices of energy resources and, in turn, fertilizers continue to rise.

RESEARCH TO PERMIT TOLERANCE OF THE STRESSES

In recent years, research has concentrated on tailoring the rice plant to tolerate or resist the physical and biological stresses in the tropics (IRRI, 1976). This focus accommodates the low-income farmer's financial inability to purchase inputs. The focus also recognizes that in some situations no other sensible alternative exists. Practically speaking, we cannot change the climate or prevent floods but we think we can provide the farmer with rices that will tolerate these adverse environments.

New rice varieties tailored to resist or tolerate stress are being developed and tested through an interdisciplinary and international "genetic evaluation and utilization" program sponsored by IRRI.

The primary components of this program are:

1. The collection and storage of seeds of rice varieties from throughout the world.
2. The systematic screening of these accessions for their ability to resist or tolerate the stresses faced by rice in the field.
3. Hybridization of appropriate parents to incorporate the desired traits into varieties known to be high yielding and to have acceptable eating qualities.
4. The international evaluation of the progeny to be certain the desired traits have been incorporated and retained, and
5. Release, by national program scientists, of the superior lines as varieties.

Germplasm Collection and Conservation

In cooperation with national program scientists IRRI has collected almost 50,000 rice accessions from every important rice-growing country. Facilities are available to store an additional 50,000 samples and an active program is under way to collect these rices. Scientists from national programs can draw seeds from the germplasm bank. Each year, IRRI sends 5,000 to 10,000 packets of seeds from the "bank" to requesting cooperators.

Systematic Screening of Accessions and Breeding Materials

Teams of "problem area" scientists and plant breeders systematically screen rices from the germplasm bank and from the Institute's breeding program for resistance to or tolerance for:

1. insects
2. diseases
3. drought
4. adverse soils
5. floods and deep water, and
6. temperature extremes.

In each case, specific methods have been developed that permit the rice plants to be placed in the presence of the stress. The ability of each accession or line to tolerate the stress is recorded. Each year tens of thousands of rice cultivars and breeding lines are evaluated. Most of the screening is in greenhouses or screenhouses, but screening for drought resistance and final evaluation for pest resistance is in the field.

Table 8 indicates the number of entries screened each year for each pest. Similar numbers are screened for drought resistance and for tolerance of adverse soils, although much of this work is done in cooperation with national program scientists.

Table 8. Screening for resistance to major rice diseases and insect
pests in 1976 at IRRI.

Pest	Entries from		Total
	Germplasm bank	Breeding lines	
Diseases			
Blast	2,631	75,000	77,631
Bacterial blight	5,712	50,000	55,712
Tungro virus	4,848	44,500	49,348
Grassy stunt virus	513	7,057	7,570
Sheath blight	7,839	15,000	22.839
Insects			
Green leafhopper	4,773	980	5,753
Whitebacked planthopper	4,009	129	4,138
Brown planthopper			
Biotype 1	7,548	21,800	29,348
Biotype 2	6,696	17,737	24,433
Biotype 3	6,163	2,051	8,214
Stem borer			
Striped borer	0	1,080	1,080
Yellow borer	0	1,034	1,034
Whorl maggot	8,557	128	8,685
TOTAL	59,289	236,496	295,785

Source: IRRI, 1977.

Hybridization

The plant breeders have the responsibility for making crosses
to incorporate the desired characteristics into superior varieties.
They make about 5,000 crosses each year (Figure 3). The progeny of
F_2 and succeeding generations are fed back through the screening
process to be certain the desired characters were in fact incorporated. F_2 and F_3 seeds are also sent to national cooperators to
evaluate and select for adaptation to local conditions.

International Evaluation

As soon as it is practical, cooperators in other countries in
Asia, Africa, and Latin America are provided seeds of promising lines
and varieties from the IRRI GEU program. This is done through infor-
mal arrangements described in the previous section and through the
formally organized International Rice Testing Program (IRTP). This
program permits the evaluation at hundreds of locations of not only

IRRI lines and germplasm accessions but similar lines and varieties
"nominated" by scientists in national programs (Table 9). In 1978
seeds for 700 such country trials were sent to scientists in 60 countries. The results of the in-country IRTP evaluation tests are submitted to IRRI, summarized, and returned to the cooperating scientists.

This international network of cooperating scientists working in
the IRTP not only provides an efficient mechanism for evaluating the
stress resistance or tolerance of breeding materials, but it helps to
identify the new strains or biotypes of the diseases or insect. The
GEU and its IRTP component are, in effect, a worldwide factory to
identify genetic sources of resistance to different stresses that
adversely affect rice yields.

The perceptions of 35 Asian plant breeders of the relative importance of different stresses on rice yields, along with the percentages
of their crosses made to accommodate those stresses, are in Table 10.
The breeders are aware of the constraints caused by pests and are
making crosses to remove them. Likewise, where deep water or low temperature problems exist the plant breeders seem to be trying to develop varieties to overcome these stresses. But for drought, adverse
soils, flood damage, and monsoon cloudiness there appears to be far
less effort to produce suitable varieties than the perceived importance of these environmental stresses would suggest. The absence of
reliable methods to measure the resistance or tolerance of the rices
to these adverse environmental conditions probably accounts at least
partly for this unfortunate situation.

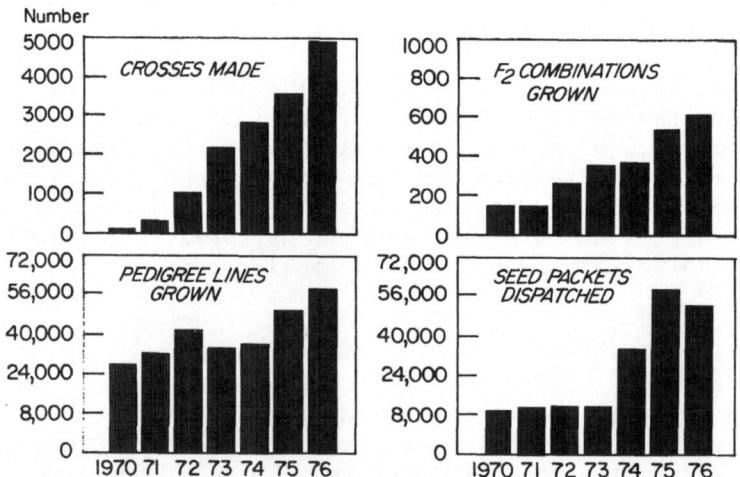

Fig. 3.   Genetic evaluation and utilization activities of IRRI have
          expanded steadily since 1970.   Source: IRRI, 1977.

Table 9. Rices from throughout the world are evaluated for tolerance
          for different stresses through the International Rice
          Testing Program (IRTP).

Nursery	Sets Dispatched	Entries[a]	Packets (total)
Yield			
Lowland			
Early	56	28[b]	4,704
Medium	50	28[b]	4,200
Late	18	28[b]	1,512
Upland	46	25[b]	3,450
General Observational	97	391	37,927
Environmental Stress			
Upland	65	153[c]	19,890
Salinity-alkalinity	35	77	2,695
Deep water	26	70	1,820
Cold	11	380	4,180
Biological Stress			
Blast	48	476	22,848
Sheath blight	32	154	4,928
Tungro	15	187	2,805
Brown planthopper	28	128	3,584
Gall midge	24	119	2,856
Stem borer	26	82	2,132
Total	577		119,531

[a]Including check varieties. [b]Three replications. [c]Two sets for
two planting dates.

RESEARCH TO SUPPORT THE "FACTORY"

Most of what has been described in this paper are applied
research and development activities, designed to answer questions
such as "how do different rices respond to different stresses?" But
we also need answers to why these responses occur. Scientists in the
tropics are attempting to find some of the answers but their capabilities and resources are limited. They are forced to turn to others
in the more developed countries who have more basic research experience, more sophisticated research equipment, and larger scientific
staffs.

Table 10. Plant breeders' perception of factors constraining rice
           yields in relation to the crosses they made to remove
           these constraints.[a]

Biological or environmental factor	% of breeders listing stress as major constraint	% of crosses made by breeders to remove constraints
Diseases and insects	100	62
Drought	85	6
Adverse soils	60	8
Monsoon cloudiness	48	0
Floods	36	4
Low temperature	30	11
Deep water	12	9
High temperature	9	0
Waterlogged soils	3	2
Others	6	0

[a]Data from 35 plant breeders in 27 experiment stations in 10 Asian
countries. Source: Hargrove, 1978.

Genetics

     One limitation to progress in the GEU program is lack of infor-
mation on the gene source of resistance to or tolerance for the
stresses. Some progress has been made in identifying the simple
genetic bases for resistance to insects and diseases such as the
brown planthopper and bacterial blight. But little is known of the
genetics of factors such as drought resistance, tolerance for adverse
soils, or protein levels.

     GEU scientists often have difficulty incorporating a certain
trait because of the seeming breeding incompatibility of the parents
involved. For example, there appears to be only one or two genetic
sources of dwarfism in modern indica rices but several such sources
have been found in japonica types. Unfortunately, incompatibility
problems have prevented the incorporation of the "dwarfing" genes from
the japonicas into indica types. Help is needed from geneticists to
make these crosses if the genetic base of semidwarf type rices of the
tropics is to be broadened.

     Plant cell tissue culture techniques have the potential to help
plant breeders who seek genetic resistance to plant stresses
(Scowcroft, 1977; Cocking and Power, 1979). While it may be a generation before new varieties can be successfully produced with these
techniques, they should be helpful to the plant breeder in screening

for desired characteristics such as the stresses being discussed in
this paper (NRC, 1977).

Diseases and Insects

One significant research priority for outside assistance on
pest management is to ascertain why certain rice varieties are resistant to these pests. Some IRRI work has clearly identified the biochemical bases for differences in resistance to the stem borer
(*Chilo suppresalis*) (Figure 4) and to the brown planthopper
(*Nilaparvata lugens*). When an extract of the stem borer-resistant
TKM6 was sprayed on the leaves of a susceptible variety, Rexoro, egg
laying by the pest was markedly reduced. Similarly, when a Rexoro
extract was sprayed on the leaves of TKM6 the egg laying was increased. A cooperating Japanese scientist, K. Sogawa, has found that
an extract of a brown planthopper-resistant variety, Mudgo, contains
an antifeedant compound that at least partly accounts for its host
resistance.

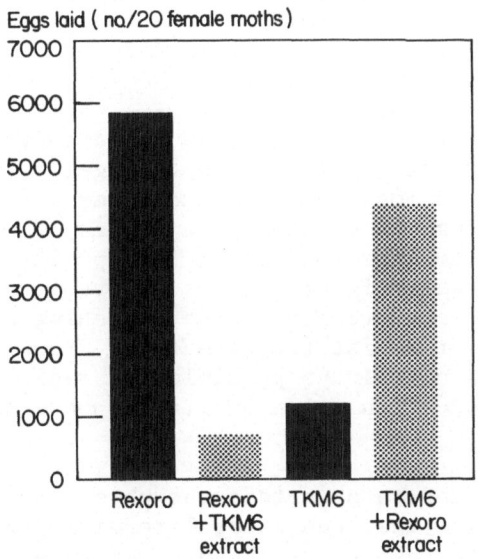

Fig. 4. Application of a plant extract from plants resistant or susceptible to the striped borer markedly changes the borer's egg laying
preference. Resistant varieties become susceptible and susceptible
varieties gain resistance to egg laying. Source: IRRI, 1978.

It is critical that this type of information be obtained on other host-pest relationships. It could be useful in the development of rapid screening techniques for host resistance, and it could also provide the basis for alternate control methods to supplement those based on host resistance.

A second basic research priority area is biological control of pests. The emergence of several biotypes of the brown planthopper, along with evolving strains of the blast fungus, calls attention to the danger of relying solely on host resistance in rice pest management. Insect predators, parasites, and pathogens should also receive attention. Some of this work can best be done in the tropics but other work should be done in carefully controlled laboratories elsewhere. IRRI scientists are encouraged by the interest of Boyce Thompson Institute scientists in studying pathogens affecting the brown planthopper, one of Asia's most serious rice pests.

Drought

The complexity of the factors responsible for drought resistance makes difficult the development of methods to evaluate resistance differences among rice cultivars. The most help is needed in the development of reliable screening methodologies. The ability to escape or avoid moisture stress, or to tolerate it, are components of drought resistance. Also in a practical sense, a cultivar's ability to quickly recover may determine its utility in low-moisture situations. All of these factors must be considered in ascertaining the drought resistance of a variety.

Research is under way to determine the effects of moisture stress on different rice varieties at different stages of plant growth. Like other cereals, rice is most sensitive to water stress at the flowering stage, sterility being one of the consequences of drought (Yoshida, 1978). But we need to know more about the specific chemical and physical processes that moisture stress adversely affects. And methods must be developed to permit the screening of rices for their ability to tolerate or resist the effects of drought at flowering. Unfortunately, it is not always possible to predict the response of a plant at the reproduction stage by its response at the vegetative stage.

Research is needed to ascertain the physiological, morphological, and biochemical traits that best correlate with drought resistance or one of its components (escape, avoidance, or tolerance). We have learned that the size, density, and distribution of roots is correlated with the drought resistance of different rice varieties; so is the presence of epicuticular wax on leaf surfaces (IRRI, 1977). However, since drought resistance is probably due to a number of seemingly unrelated characters, it is difficult to identify the characters or combination of characters that governs the degree of drought resistance in a given variety. The struggle to clearly identify characters that can be used to screen varieties for drought resistance is formidable and will need the help of the best minds in developed as well as developing countries. While much of the work must be done in the tropics, some can and should be done under controlled conditions in the more developed countries.

Adverse Soils

A better understanding is being sought as to how certain cultivars are able to tolerate soil toxicities such as extreme acidity, alkalinity, or salinity. We need to know the physiological bases of differences in tolerance among varieties. For example, what are the mechanisms by which the variety Pokkali from India is able to tolerate high iron levels of very acid soils, and also high salinity levels, which are more often associated with soils of high pH. Knowledge of the mechanism of resistance of Pokkali and of other varieties to adverse soil conditions should be helpful to the soil chemist and to the plant breeder in developing screening techniques, and in producing varieties tolerant of adverse soils.

Except for Japan, the more developed countries have done little research on the nitrogen cycle in wetlands. Most of their attention, including that concerned with biological nitrogen fixation, has been oriented to dryland soils. The anaerobic environment in submerged soils must be studied along with nitrogen changes that occur in this environment to determine if nitrogen utilization efficiency can be increased in these wetland soils.

We also need to learn the practical significance of biological nitrogen fixation by free living organisms in the rhizosphere, or by blue-green algae or the azolla-anabaena complex. Recent findings of Boyce Thompson Institute scientists stationed at IRRI offer clear proof through direct experimentation that significant quantities of nitrogen can be fixed in the rhizosphere of the rice plant. This finding concurs with earlier research using acetylene reduction techniques. Research is now needed to ascertain factors that influence the rhizospheric fixation process.

The practical potential of the azolla-anabaena complex must be further explored. Chinese and Vietnamese farmers use azolla in crop rotations and consider it an excellent source of nitrogen. Basic research is needed, however, to ascertain if there are differences among strains of azolla in the amount of nitrogen that they can fix. IRRI has been requested to collect azolla germplasm but the expense of maintaining large numbers of vegetative samples of this aquatic fern makes large-scale collection impractical. Extensive germplasm collection probably must wait until a practical method of inducing

sporulation of the azolla is developed. Basic researchers must be
oriented to accomplish this task.

Flood Tolerance

As with the other stress areas, scientists need to better understand why cultivars differ in their tolerance of floods and of deepwater conditions. What triggers stem elongation in one variety as
floodwaters rise, permitting the plant to survive, while another
variety does not elongate? Likewise, what is the physiological
basis for differences in tolerance for submerged conditions? Why
can one cultivar remain submerged for a week with little apparent
damage while another "drowns out"?

Finding answers to these questions would probably interest a
basic researcher purely on the basis of scientific curiosity. The
same answers can be used by IRRI scientists and cooperators in
national programs to develop screening methods for these important
traits, thereby accelerating the rate of development of improved
varieties for flood-prone and deepwater areas.

SYMBIOTIC ALLIANCES

The development of varieties and technologies to adjust to or
remove environmental and biological stresses is one of the most critical scientific as well as socioeconomic needs of the humid tropics.
Alliances must be formed between scientists located in the tropics
and those in the more affluent temperate countries. Scientists in
Japan, the U.S., Europe, and Australia-New Zealand have the capability and the resources to get answers to the "why" questions that
have been posed here. They can help obtain an understanding of the
biochemical and physiological bases for differences among cultivars
in their response to stress. Applied scientists in the tropics can
use the information they obtain to develop improved methods to measure
plant response to stress. In turn, this information can become the
basis for the accelerated screening of cultivars for resistance to
or tolerance for the stresses and for the development of alternative
means of removing or accommodating the stress.

REFERENCES

ADB (Asian Development Bank),1977, "Rural Asia, Challenge and Oppor-
    tunity," Federal Publications, Singapore.
Aziz, S., 1977, The World Food Situation - Today and in the Year
    2000, in: "Proceedings of the World Food Conference of 1976,
    June 27-July 1, Iowa State University, Ames, Iowa, USA," Iowa
    State University Press, Ames, Iowa, USA.

Blair, G. J., Mamaril, C. P., and Momat, E., 1978, Sulfur Nutrition
   of Wetland Rice, IRRI Res. Pap. Ser., 21.
Bradfield, R., 1970, Multiple Cropping in the Tropics, in: "Research
   for the World Food Crisis," American Association for the
   Advancement of Science, Washington, USA.
Brown, L. R., 1978, "The Twenty Ninth Day: Accommodating Human
   Needs and Numbers to the Earth's Resources," W. W. Norton and
   Co., Inc., New York, New York, USA.
Chang, T. T., and Oka, H. I., 1976, Genetic Variousness in the
   Climatic Adaptation of Rice Cultivars, in: "Climate and Rice,"
   International Rice Research Institute, Los Baños, Philippines.
Cocking, E. C., and Power, J. B., 1979, Application of Tissue
   Culture to Plant Breeding, in: "Plant Breeding Perspective,"
   J. Sneep and A. J. T. Hendriksen, eds., Pudoc, Wageningen.
Dalrymple, D. G., 1978, Development and Spread of High-Yielding
   Varieties of Wheat and Rice in the Less Developed Nations,
   For. Agric. Econ. Rep., 95, U.S. Department of Agriculture
   and USAID, Washington, D.C., USA.
Eckholm, E. P., 1976, "Losing Ground: Environmental Stress and
   World Food Prospects," W. W. Norton and Co., Inc., New York,
   New York, USA.
FAO (Food and Agriculture Organization), 1976, Monthly Bull. Agric.
   Econ., 25(3).
Hargrove, T. R., 1978, Diffusion and Adoption of Genetic Material
   Among Rice Breeding Programs in Asia, IRRI Res. Pap. Ser., 18.
IRRI (International Rice Research Institute), 1975, "IRRI Research
   Highlights for 1974," International Rice Research Institute,
   Los Baños, Philippines.
IRRI (International Rice Research Institute), 1976, "World Rice
   Statistics," Department of Agricultural Economics, Interna-
   tional Rice Research Institute, Los Baños, Philippines.
IRRI (International Rice Research Institute), 1977, "IRRI Research
   Highlights for 1976," International Rice Research Institute,
   Los Baños, Philippines.
IRRI (International Rice Research Institute), 1978, "IRRI Research
   Highlights for 1977," International Rice Research Institute,
   Los Baños, Philippines.
IRRI (International Rice Research Institute), 1978a, "Summary Report
   on the First and Second International Trials on Nitrogen Fertil-
   izer Efficiency in Rice (1975-1977)," Revised, International
   Rice Research Institute, Los Baños, Philippines.
IRRI (International Rice Research Institute), 1979, "Long Range
   Planning Committee Report," International Rice Research
   Institute, Los Baños, Philippines.
Isrankura, V., 1966, "A Study on Rice Production and Consumption in
   Thailand," Ministry of Agriculture, Bangkok, Thailand.
Jennings, P. R., 1976, "The Amplification of Agricultural Produc-
   tion," Sci. Am., 235(3):180-190.
Knight, C. G., and Wilcox, R. P., 1976, "Triumph or Triage: the
   World Food Problem in Geographical Perspective," Association

of American Geographers, Washington, D.C. Res. Pap. 75-3.

National Academy of Sciences, 1975, "World Food and Nutrition Study: Enhancement of Food Production for the United States," Report of the Board on Agriculture and Renewable Resources Commission on National Research Council, National Academy of Sciences, Washington, D.C.

National Research Council, 1977, "World Food and Nutrition Study: the Potential Contributions of Research," National Academy of Sciences, Washington, D.C.

Ponnamperuma, F. N., 1977, "Soil Factors Affecting Plant Stress," Talk given at the International Conference on Stress Physiology of Plants for Food Production, Boyce Thompson Institute for Plant Research, New York.

Sanchez, P. A., and Buol, S. W., 1975, Soils of the Tropics and the World Food Crisis, in: "Food: Politics, Economics, Nutrition, and Research," P. H. Ableson, ed., American Association for the Advancement of Science, Washington, D.C.

Scowcroft, W. R., 1977, Somatic Cell Genetics and Plant Improvement, Adv. Agron., 29:39-81.

Spain, J. M., Francis, C. A., Howeler, R. H., and Calvo, F., 1975, Differential Species and Varietal Tolerance to Soil Acidity in Tropical Crops and Pastures, in: "Soil Management in Tropical America," E. Bornemisza, ed., North Carolina State University, Raleigh, North Carolina.

Takase, K., and Wickham, T., 1976, "Irrigation Management as a Pivot of Agricultural Development in Asia: Report by the Associate Experts in Irrigation for the Asian Agricultural Survey II," Asian Development Bank, Philippines.

USDA (United States Department of Agriculture), 1974, The World Food Situation and Prospects to 1985, For. Agric. Econ. Rep., 98, USDA Economic Research Services, Washington, D.C.

Wortman, S., and Cummings, R. W., Jr., 1978, "To Feed this Hungry World The Challenge and the Strategy," The Johns Hopkins University Press, Baltimore and London.

Yoshida, S., 1978, Tropical Climate and Its Influence on Rice, IRRI Res. Pap. Ser., 20.

CROP IMPROVEMENT

John K. Coulter
Scientific Adviser
Consultative Group on International Agricultural Research
Secretariat
World Bank
Washington, D.C.

INTRODUCTION

In this discussion on crop improvement I shall be describing
some of the work of an organization, the Consultative Group on
International Agricultural Research, that spends about 60% of its
resources (approximately $100 million in 1979) on the improvement
of 13 or 14 food crop species that are particularly important in
the developing world. To contribute to what I understand is the
theme of this meeting, the interaction between basic -- in my
definition mission-oriented or strategic research, i.e., generation
of scientific ideas in which advances in agricultural technology
can be based -- and technology development, I will try to describe
what is being done in technology generation in the international
centers. From this I will attempt to define some of the questions
that these technology generation programs raise and hopefully
indicate ways in which more productive linkages could be created.

THE NEED FOR IMPROVED TECHNOLOGY

The need for improved technology for agriculture in the
developed countries requires no emphasis, but by way of background
there may be a few aspects that bear special attention. High
priority is being given to increasing food production as a means
of improving the lot of the rural poor, including the small landholders, tenants and the landless. It should be emphasized that
the gains from increased food production accrue both to those who
produce food and to those who consume it. Low-income consumers

may typically spend more than 80% of their income on food, so any
reduction in cost benefits them disproportionately.

In the past, increasing the area committed to crops has played
a major role in increasing production. This expansion has occurred
both through cultivation of new lands and through intensifying production on existing land by shortening the fallow period; a short-
term gain, but probably a long-term loss. Growing pressure on
land in the past couple of decades has meant that about two-thirds
of the increased production in Asia and the Middle East/North Africa
regions has come from enhanced productivity. However, there is very
little unused land in regions where food needs are greatest, since
more than 80% of the potentially-arable land is already being used
in Asia and North Africa/Middle East. This is a theme to which I
wish to return later -- that when we are discussing improvement of
agricultural technology, we are talking not only of improvement in
the better endowed areas, but also of many areas that are marginal
from both the soils and climatic point of view.

This seems particularly important in the context of discussions
on basic research. In the U.S., for example, marginal land --
and here marginal can have a range of definitions -- can go out of
agricultural production, as people find alternative employment.
Furthermore, where marginal lands are being farmed in the U.S.,
the farmers, by and large, have greater protection against risk,
at least that ultimate of risks -- starvation.

The magnitude of the task of improving food production is such
that output in the developing countries would need to increase by
about 4% per annum compared with an actual average increase of about
2.7% between 1960 and 1975. However, a growth rate of 4% per annum
has not been achieved countrywide anywhere on a sustained basis.
Thus, such an increase will need not only the cultivation of more
land, but also greatly increased inputs of labor, water, fertilizers,
power, improved seeds and better infrastructure like roads and
markets to supply these inputs. The unique role of agricultural
research is that of generating the technology that will make these
inputs more efficient.

The poor record of yield improvement in food crops in the
tropics is often attributed to lack of investment in agricultural
research in those crops. This is probably correct, although it
should also be emphasized that the quality of the research on cash
crops has usually been better because they have had more stable
long-term funding, and because the producers have often constituted
a more articulate pressure group insisting that the research funds
be used effectively.

The disappointing record in yield improvement of food crops,
and the apparent ineffectiveness of much of the research led to the

creation of the first of the international centers, the International
Rice Research Institute (IRRI), in 1960. This was followed in 1966
by the International Maize and Wheat Improvement Center in Mexico.
The two centers had a narrow focus on specific crops of critical
importance to world food supply. Within six years of its creation,
IRRI had developed the first tropical ecotype of a fertilizer-
responsive, short-strawed rice variety. The Rockefeller program,
the predecessor of the International Maize and Wheat Improvement
Center, had earlier developed similarly successful dwarf varieties
of wheat.

The Consultative Group on International Agricultural Research
(CGIAR) was organized in 1971 to broaden financial support and ensure
long-term funding for the international programs of research and
training. It was established as an association of international
and regional organizations, national governments, public and private
foundations and representatives of developing countries. There are
currently over 5,000 staff, including 400 international scientists,
working at the centers supported by the Group.

We now ask the question -- what do we get for our investment
in this system? I want to examine this in two ways, one very
briefly, the impact on production in the farmers' field, the other
the main theme of my talk -- the output of the system in terms of
scientific knowledge and technology development.

Some aspects of the former are illustrated by tables I and II
showing the rate of adoption of modern varieties of wheat in various
countries, and the yield of modern varieties of rice in the
Philippines. These tables demonstrate two important features, the
rapid rate with which modern varieties can spread and the relatively
modest yields being obtained from the new rice varieties in the
farmers' fields.

Table I. RATES OF ADOPTION OF NEW VARIETIES
          (In Percentages of Total Area under Wheat)

Year	Afghanistan	India	Nepal	Pakistan	Tunisia
66/67	–	4	5	2	–
67/68	1	20	13	16	–
68/69	4	30	36	38	2
69/70	5	30	34	43	7
70/71	9	36	43	52	11
71/72	10	41	52	57	6
72/73	15	51	66	56	9
73/74	16	57	76	59	6
74/75	17	62	85	63	6

Table II: PADI YIELDS (KG/HA) - REPUBLIC OF THE PHILIPPINES

Crop Year	Irrigated			Rainfed (Lowland)		
	Modern Varieties	Other	Area in % Modern Var.	Modern Varieties	Other	Area in % Modern Var.
67/68	1,967	1,613	34	1,307	1,239	17
68/69	1,778	1,617	62	1,125	1,089	31
69/70	2,155	1,886	61	1,487	1,527	39
70/71	2,023	1,930	67	1,614	1,580	45
71/72	2,053	1,723	73	1,443	1,350	55
72/73	1,950	1,741	70	1,276	1,110	60
73/74	2,051	1,887	80	1,531	1,252	64
74/75	2,222	1,879	79	1,430	1,179	64

Source: Bureau of Agricultural Economics, Department of Agriculture,
          Republic of the Philippines.

TECHNOLOGY DEVELOPMENT

The international centers are involved in a number of aspects
of technology development, but about 60% of their efforts go to
plant improvement. The remainder is spent on various kinds of
farming-systems research and on animal sciences.

The CGIAR system is the largest coordinated effort in biological research in the developing world. As will be discussed
later, its global links enable it to accelerate research processes.
Whilst such research creates major benefits for the developing
countries in providing new technology, in training scientists and
in providing an intellectual environment that is conducive to productive research, the scientific community and ultimately the
farmers of the developed countries will also receive substantial
benefits.

In this paper I would like to give some attention, not only
to technology development by the centers, but also to ways in
which the centers have contributed to the exploitation of existing
scientific ideas by institutional innovation and to the centers'
influence on the kinds of goals that are set by research organizations in the developing countries. An example of the latter can
be drawn from the changing goals of the individual centers. The
goals of the early IRRI and CIMMYT programs were the production
of plant types that would be highly responsive to fertilizers,
daylength insensitive, and resistant to some of the major diseases;
in other words high productivity. This goal still remains.
Nevertheless, the studies of the impact of the new technology also

revealed that output would not increase significantly in many
areas unless the resource-poor farmers could also participate in
the benefits of more productive technology. Such technologies
must not require resources that the farmers are unable or unwilling
to put into this production.

The explicit recognition of these limitations has led to
changes in breeding programs, to the recognition of the importance
of studying existing farming systems, and consequently to greater
emphasis on yield stability. This change in goals, at variance
with the philosophy of many scientists, has had considerable
impact on the goals of some national research programs, but there
are others which strongly dispute it on the argument that to
follow this strategy is to condemn the farmer to a low-productivity
economy.

No country can afford to neglect regions with high production
potential, if only to produce cheaper food for the urban population. The variation in soils and climate influence greatly the
distribution of crops for there is obviously a strong correlation
between crop species and environment. Bread wheats, for example,
are grown in the better rainfall areas, barley is the crop of poor
farmers on poor land; sometimes cash crops, e.g., cotton, may
occupy the better land, food crops, like cassava, the poorer areas.

The introduction of global testing is an example of a vast
expansion in scale of an idea that began many years ago. Now there
are several testing networks operated by all the centers working
on plant improvement. These may cover as many as 80 or 90 countries
with several sites in each country (Table III). Global cooperation
on protection against serious pests is also particularly important
in an age when these can move rapidly not only from one tropical
country to another, but from the tropics to the temperate countries.
Thus, the spread of cassava mealy bug from South America to Africa,
illustrate the dangers, the need for such cooperation and the role
that the international centers can play in preparing to meet such
problems.

CONSERVATION AND UTILIZATION OF GENETIC RESOURCES

Since he began cultivating plants about 10,000 years ago, man
has made great use of the genetic variation in the relatively few
species which he used for food and clothing. Although genetic
diversity can be created by mutation breeding, by far the largest
reservoir of such variation (gene pools) remains in the many
cultivated varieties and closely related wild species. The four
basic forms of germ plasm can be summarized as:

   (1) breeding material and varieties currently in use and
       usually freely interchanged;

Table III: INTERNATIONAL TESTING OF SOME MAJOR CROPS BY IARCs
             (1978).

Crop	Number of Crosses/ Pollinations	Nurseries/Lines	
		Number Sent	Number of Countries
Rice	5,000	577	41
Bread Wheat	10,000	595	95*
Durum Wheat	5,000	352	56*
Triticale	4,000	400	77*
Maize	17,000**	3,470	80*
Barley	2,500	258	74*
Cassava (CIAT)	500	30	20
Chick-pea	1,200	327	28
Pigeon Pea	500	39	20
Cowpea	100,000**	39	20
Bean	1,500	250	30
Groundnut	50,000**		

* These include some nurseries in developed countries.
**Pollinations.

(2) cultivars that have become obsolete but are still stored
    in national collections (developed countries have many of
    these, developing countries often rather few);

(3) primitive cultivars and land races or varieties that
    have been used, often for centuries, in the lesser
    developed countries;

(4) related wild species that may carry certain important
    characters.

The threat to these irreplaceable resources of primitive
cultivars and related wild species has become increasingly recognized during the past two or three decades, and the International
Board for Plant Genetic Resources was set up in 1974 to organize
cooperative efforts on this problem. The Board has been active
in organizing collecting expeditions, especially of threatened
species or in vulnerable areas; in proposing plans for storage
of such material; in drawing up of agreed lists of descriptors
for the establishment of suitable, computerized information storage
and retrieval systems; and in training scientists for work in
genetic-resources conservation. The long-term impact of thes.
operations will benefit both the developed and developing countries,
for several of the species collected (e.g., potatoes, maize, sorghum,

food and fodder legumes) are important crops in developed countries,
too. The magnitude of this enterprise at the International Agricultural Research Centers is demonstrated by the information in
Table IV.

CROP IMPROVEMENT

     Plant breeding has been likened to a factory in which raw
materials, i.e., genetic variation, are passed through a series of
processes with improved varieties as the end product. In the
international centers, material may be taken off the "assembly
line" at various stages, and passed on to national programs for
finishing according to their particular needs; or finished varieties
may be produced by the center. Since plant breeding is basically
a system of managing the direction and increasing the speed of the
processes that occur in nature, it becomes mainly a question of
manipulating large numbers of plants, as combinations of desirable
characters take place by chance.

     The plant breeding techniques of the centers are the same as
those used by national programs throughout the world, but they do

Table IV: CENTER GERM PLASM COLLECTIONS

Crop	Number of Accessions	Center
Maize	13,000	CIMMYT
Pearl Millet	5,500	ICRISAT
Rice	37,000	IRRI
Sorghum	15,000	ICRISAT
Cassava	2,500	CIAT/IITA
Potato	12,000	CIP
Cowpea	8,000	IITA
Chick-pea	11,300	ICRISAT/IITA
Phaseoleus sp.	21,000	CIAT
Pigeon Pea	6,000	ICRISAT
Soybean	800	IITA
Groundnut	7,000	ICRISAT
Forage legumes	3,500	CIAT

differ in capacity in a number of ways. The centers have more
genetic variation at their disposal by way of germ plasm collections
than do most national programs. They have the staff and the organization to handle a greater number of crosses (Table III). They have
been able to double the number of generations and thus half the
time by "shuttle" breeding, i.e., carrying out the crossing programs twice a year by operating in two different climatic regions.
Finally, they have been able to use the extensive international
testing networks in which centers and national programs exchange
and test germ plasm. The success of such networks obviously depends
on the quality of the work done in each country program and the
timeliness with which returns are made. There are considerable
variations in these from country to country.

OBJECTIVES IN CROP IMPROVEMENT AT THE CENTERS

     The ultimate objective in plant breeding is higher productivity.
It may be higher productivity per unit of input, per unit of land,
or per man-day. A national program operating for a particular
region of a country may have very specific objectives, such as an
answer to a distinctive pest problem, a soil deficiency, a climatic
characteristic or a consumer preference. Because of their mandates,
the international centers have to produce materials that have the
inherent ability to perform well under a very wide range of conditions. This had led to much emphasis on the concept of wide
adaptability. The centers are also concerned that their materials
are also stable in yield from year to year. Even for locale-
specific materials, plant breeders aim at varieties that will give
relatively superior yield each year and this requires testing for a
number of years before release. By testing materials widely at
many sites, the pest and climatic variations that might appear
over several years at one or a few sites may be encountered in one
year. Thus materials which show adaptability, should also show
stability. Though the theory is not proven, there seems enough
evidence to indicate that it is a valid approach. Certainly in
irrigated rice and wheat, the wide adaptability that is necessary
to cope with a range of pest problems has been evident. Whether
such wide adaptability will be achieved in crops for rainfed agriculture, is another question. The answer depends on the magnitude
of the genotype X environment interaction. Increased productivity
in dryland agriculture will depend on such factors as timeliness
of cultivation (more difficult for the small farmer to control) as
well as on improved varieties. The needs of rainfed agriculture
will also require more attention to such locale-specific traits as
early sowing, early or late flowering, drought tolerance, and
pest tolerance.

     Another aspect of yield stability concerns the breeding of
varieties which will perform at least as well at low or nil inputs

as the traditional varieties, and yet perform more efficiently
when inputs are available. Pest resistance must be as good in the
new as in the old varieties and, in general, this has been the case.

Because concern with the resource-poor farmer has dominated
much of the thinking of the Consultative Group, there is now a
concerted effort by the international centers to design plant
varieties and farming techniques that will have low requirements
for purchased inputs, and thus partially eliminate the need for the
sophisticated infrastructure to make these inputs available. One
aspect of this is the attention that has been devoted to pedigree
breeding and production of synthetics rather than hybrid production.
So far this century, production in three major grain crops, maize,
sorghum and pearl millet, has been revolutionized by hybrid production. Hybrid vigor in sorghum, for example, confers resilience
in the plant to better withstand moisture stress. Hybrid wheat
has had considerable attention by CIMMYT, although some so far
insuperable problems have been encountered. Finally, IRRI has done
some research on hybrid rice. Work in China suggests that hybrid
rice may yield 10-30% more than the normal varieties. In spite
of the advantages and the fact that there are examples where hybrids
have been successful in the developing countries, e.g., hybrid
maize in Kenya, the general consensus seems to be that the requisite
delivery system to make hybrid seed available does not normally
exist.

About 20% of the plant improvement programs at the centers is
devoted to grain and legumes. In general, it is more difficult
and more time-consuming to achieve improvements in these. Table V
shows they suffer from a wide range of pests. In some there is a
marked dormancy period which prolongs the time between generations.
The kind of dramatic yield increases that have been obtained in
rice and wheat are unlikely to appear in these crops.

Another aspect of maximizing yields occurs in the context of
raising yield ceilings. The difference between the yields attained
on experimental plots and those on farmers' fields is so great that
many scientists maintain that most of the research effort should
go towards closing this gap rather than to raising the potential
ceiling. Nevertheless, from the point of view of the scientist,
achieving record yields can be very important since it is only by
such means that the genetic potential can be defined. In trying
to reach such yields, breeders can define the components of existing
varieties which need to be altered in order to increase the yield
potential.

Finally, it must be emphasized that whilst improvements in
yield stability can be obtained through better pest resistance and
ability to withstand drought, the opportunities for increasing

yields in the absence of better farming practices and increased
inputs are limited.

PEST RESISTANCE

In the tropics and sub-tropics, the range and severity of
pests are more extreme than they are in temperate zones, and less
is known about them. Moreover, they tend to be more persistent
due to the absence of the natural "break" caused in temperate
climates by winter cold; in some areas year-round cropping provides
the host plants to sustain the pest. Table V shows some of the
major pests of crops included in center mandates. It should be
recognized that these vary in importance from year to year and region
to region. Furthermore, a once relatively minor pest can assume
major importance very rapidly. Examples are the brown planthopper,
a minor pest in the 1960's, but one of the most serious today, and
tungro virus, which is a very serious disease in rice. These
changes cannot be predicted until substantial areas of new varieties
are grown by farmers, i.e., they are second-generation technical
problems and thus rice and wheat are the only center crops on which
such information is available.

It is sometimes suggested that the introduction of modern
varieties has left the farmers more vulnerable to pest ravages
than when they grew traditional varieties. The evidence does not
support this. In fact, many varieties became obsolete because of
disease susceptibility. However, large areas planted to one
variety obviously present a threat of dramatic damage if a new
pest occurs. It has been estimated that 25% of the 130 million
ha of the world's rice is in short-strawed, high-tillering photo-
period-insensitive varieties. Furthermore, it is calculated that
this area of some 32 million ha is planted with only about 100
varieties which have replaced hundreds of the traditional cultivars.
In Korea, for example, the variety Tongil, first released in 1972,
covered nearly 40% of the total area by 1977. So far, the availability of a wide range of genetic materials has enabled the centers
to identify sources of resistance to the majority of these pests
so that tolerant or resistant varieties have either been developed
or are being tested. For some diseases a range of resistant
materials has been found, but in others, very few sources of
resistance exist. In grassy stunt disease in rice, for example,
out of nearly 7,000 accessions of cultivated rice and several
wild species of Oryza evaluated, only one accession of Oryza nivara
was found to be highly resistant. The search for resistance to
blast (Pyricularia oryzae) in rice illustrates the enormous amount
of work that has been needed to develop varietal stability to this
highly variable organism. The disease now appears reasonably well
under control in lowland rice in Asia, but remains a major threat
to upland rice, especially in Africa.

Table V: SOME MAJOR DISEASES AND PESTS OF CENTER CROPS

Crop	Disease	Pest
Rice	Blast, Bacterial Blight, Tungro, Grassy Stunt, Ragged Stunt	Brown Planthopper, Green Leafhopper, Gall Midge, Whorl Maggot, Leaf Folder
Wheat	Stripe Rust, Stem Rust, Leaf Rust, Root Rotting Fungi, Septoria, Head Scab	A number, but not as important as with maize, for example
Maize	Streak Virus, Downy Mildew, Stunt	Stemborers, Leafhoppers, Earworm
Millet	Downy Mildew, Ergot, Smut, Rust, Blast	Of minor importance in India. In Africa there are a number of serious pests, e.g., Headworm
Sorghum	Striga, Downy Mildew, Grain Mold, Charcoal Rot	Shootfly, Stemborers, Midge
Barley	Leaf Rust, Stripe Rust, Scald, Powdery Mildew, Leaf Spot, Net Blotch, Virus Stripe	A number, but not as important as with maize
Chick-pea	Wilt Complex, Stunt, Ascochyta Blight	Heliothis
Pigeon Pea	Fusarium Wilt, Sterility Mosaic, Phytophthora Blight	About 200, the most serious being Borers and Pod Fly
Field Beans	Common Mosaic Virus, Rust, Anthracnose, Angular Leaf Virus, Bacterial Blight, Golden Mosaic Virus	Leafhopper, Chrysomelid sp., Bruchid sp.
Cowpea	Leaf Spot, Anthracnose, Rust, Target Spot Bacterial Blight, Bacterial Pustule, Yellow Mosaic, Aphid-borne Mosaic, Golden-Yellow Mosaic	Leafhoppers, Thrips, Pod Borers, Aphids, Bruchid sp.
Groundnut	Rust, Leafspots, Peanut, Mottle, Rosette, Peanut Stunt, Tomato Spotted Wilt	Leaf Miners, Pod Borers, White Grubs, Aphids and Thrips
Cassava	Cassava Mosaic Disease (Africa), Bacterial Blight, Anthracnose, Super Elongation, Phoma Leaf Spot, Complex of Root Rots (Latin American and Caribbean)	Mealy Bug, Whitefly, Green and Red Spider Mites, Thrips, Scale Insects, Hornworm
Sweet Potato	S.P. Virus Disease	Weevil, Aphids, White Flies
Potato	Late Blight, Black Wart, Pink Rot, Phoma; Bacterial Wilt, Leaf Roll Virus, Virus X, Virus Y and numerous other identified and unidentified viruses	Potato Cyst Nematode, Root Knot Nematode, False Root Knot Nematode (at least 9 species)

Resistance to a number of pests is based on single dominant genes and breakdown in resistance can occur rapidly, and in some instance, disease pressure has resulted in new resistant varieties being required about every three years or less. This is by no means an uncommon pattern or one which is limited to developing countries. What is different, however, is that few developing countries have the capacity to maintain an intensive breeding program for resistance to new pests or diseases. The centers are playing a major role in helping them to develop this defensive research capability and increasing attention is being given to polygenic or general resistance.

When new pests are discovered, the centers can draw on their germ plasm collection to try to identify sources of resistance; in the search for pest-resistant genes, wild species are particularly important, as adequate resistance does not always exist in cultivated species. Exploiting these genes may be difficult and several years are required to breed other needed qualities into the crosses. Nevertheless, utilization of a comprehensive collection of genetic resources gives the plant breeder the best chance of keeping up with the evolution of new pests.

A vital element of the pest and disease resistance program is the international testing programs established by centers for crops within their mandates. With the assistance of collaborating national programs, varietal trials can be held under widely differing conditions, including known disease "hot spots", in order to detect new pathogens and to test plants against known diseases and pests. These programs benefit both centers and national research organizations.

Varietal resistance to pests is, however, only one element of an integrated control program, the others being biological and chemical control. Identifying natural predators, introducing sterile males to disrupt the mating cycle, the use of pheromones to confuse male insects, and testing various pesticides are among the techniques being investigated and tested. However, wherever possible, inbuilt plant resistance is to be preferred, because this adds little or nothing to the cost of production, whereas unfavorable weather and the high cost or unavailability of pesticides may prevent their application and in addition, add to the complexity of the technological package.

BREEDING FOR ADVERSE ENVIRONMENTS

Breeding for tolerance to heat, cold, drought, and infertile soils is also included in many center programs. Developing varieties for harsh environments presents difficulties to the plant breeder and drought tolerance presents many problems due to the many forms in which drought occurs. In wheat, useful tolerance to high-exchangeable aluminum has been discovered and in rice tolerance of

high iron levels and of salinity has been found in a number of
sources. Progress in producing more productive varieties is likely
to be slow in the programs for adverse environments, but even small
advances will be very useful in these difficult environments.

PLANT QUARANTINE

    A final aspect of a breeding program, which involves moving
thousands of packets of plant material around the world, is the
question of disease transmission and plant quarantine. Every
country is naturally fearful of the introduction of new pests,
but the standard of plant quarantine varies considerably. Some
countries have rather lax arrangements; others operate a very
stringent system which may slow down the movement of material.
The centers have developed effective seed treatment techniques
which can be operated on a large scale. Those centers dealing with
clonal material, IITA, CIP, and CIAT, have developed various tissue
culture techniques that ensure freedom from viruses. Use of true
seed rather than clonal material can also overcome some of the
problems, though even this presents difficulties, e.g., spindle
tuber virus in true seed of potatoes.

    In spite of the problems, the quantities of materials which
are moved and the general facility with which they move, indicate
that the recipient countries have a sound trust in the quality of
the centers' materials.

BASIC RESEARCH

    Much of the research at the international centers and in
national programs has depended on the scientific ideas emanating
generally from the developed countries, although there has been a
spin off of new scientific knowledge from technology development
at the centers. However, international centers and national programs have to ask the questions as to where the next advances in
productivity are to come from and what strategies should be followed
to develop the scientific ideas on which these will be based.
Against this background I would like to suggest a few areas where
mission-oriented basic research may be needed to produce the new
scientific ideas on which further advances in technology can be
made.

    In looking at the needs of the developing countries we are in
fact trying to make judgments about the possible critical areas a
decade or so hence. In doing so we can make some broad strati-
fications that might be useful in determining priority areas. In
irrigated agriculture, technologies exist already that can increase

productivity greatly. However, there are major pest problems which threaten to erode gains already made. These comprise the second generation technical problems; their solution will require both strengthened capacity in national programs, since many of the problems are locale specific, and additional research by the centers and collaborating agencies to provide the basic knowledge of genetics and of the mechanisms of tolerance and resistance. The second area concerns raising the yield ceiling which presents a challenge to the plant physiologist. The introduction of semi-dwarf varieties provided the initial breakthrough, but the irrigated areas have a secure moisture supply and often a very high level of incoming radiation so that even higher yield ceilings might be achievable.

The next broad ecological zone that could be considered is one where rainfall is reasonably reliable and where flooded rice or other rainfed crops are grown. In these regions the farmers would be expected to use inputs like fertilizers, improved planting and cultivation techniques, herbicides, etc. In countries without extensive irrigated food crop production, such areas form their major granaries. These areas thus need improved technology like better cultivation techniques, better weed control, more efficient fertilizers (including biological N fixation) and stable yielding varieties. Perhaps the major challenges here are in improved pest control, and more efficient nutrient supply.

A third kind of area is that where crop substitution is required. For various social and economic reasons, wheat is replacing coarse grains and cassava and even rice in some places, as a favored food. Hence there is a strong desire to "tropicalize" wheat. Potatoes are also under increasing demand in the tropics and we have the example of Triticale as a man-made crop which may take over from wheat in some areas. If "tropicalizing" some crops is given a high priority, a considerable amount of new research in several disciplines, particularly disease resistance, will be necessary.

Finally, there are what can be called the resource-poor areas where many of the poorest farmers and rural landless have to eke out a meagre existence. Ecologically such areas cover a very wide range of conditions -- the drought-prone areas, the steep and rocky lands, highly weathered and highly leached soils, the cold plateaus and the deeply-flooded river basins. As noted earlier, these are areas that in rich countries might go out of agriculture and into forestry, ranching or amenity areas. However, they presently support many millions of peoples for which the foreseeable technology does not appear to offer great potential. Improvements in barley and potatoes, in sorghum and millet and in minor crops like quinoa and teff may help. Commercialization of jojoba and guayule may aid some of the very dry areas. The picture is not all gloom, but it seems that these areas require more attention from scientists,

particularly a realistic assessment of what might be done by foreseeable technologies and a scientific judgment on what basic research
might achieve.

CONCLUSIONS

Many farmers and consumers in the developing countries have
benefited greatly from crop improvement; many more will benefit
in the next decade, but considerable areas may be left largely
untouched by foreseeable technologies. There are plenty of areas
of challenge for the scientist in both basic and applied research,
but determining priorities is a major task.

It is sometimes suggested that it is not funds but good ideas
which are in short supply in scientific research. As I indicated
initially, the Consultative Group has increased its investment
in research very rapidly indeed. For perfectly understandable
reasons, the first interest of the donors to this system is in the
impact in the farmer's field; therefore much of the funding will
continue to be for technology development. However, if there is a
real necessity for the system to become more deeply involved in
mission-oriented research, either directly or indirectly, then we
need to give more attention not only to what should be done but how
it should be done.

THE ROLE OF PHYSIOLOGY IN CROP IMPROVEMENT

Paul J. Kramer

Department of Botany
Duke University
Durham NC 27706

INTRODUCTION

Crop yield is controlled by the interaction between the
genetic potentialities of crop plants and the environment in
which they grow. Variations in the genotype and in the environment, including weather and cultural practices, act through
physiological processes to control growth. Thus the physiological
processes of plants are the machinery through which both the
genetic potentialities and the environment operate to produce the
quantity and quality of growth or phenotype which we term yield.
This is shown in Figure 1.

The only way in which unfavorable environmental factors
such as drought or temperature, cultural practices such as fertilization, an insect pest, or a plant pathogen can affect yield
is by affecting the physiological processes of the plant. A
drought reduces growth because plant water deficit causes
closure of stomata and reduction in photosynthesis, loss of
turgor and cessation of cell enlargement. More severe stress can
cause irreversible injury at the cell and molecular level.
Deficiencies of nitrogen and mineral nutrients reduce growth
because these substances are essential for certain physiological
processes. Defoliation by insects is injurious because it
reduces the leaf surface available for photosynthesis, and
damage to root systems reduces the uptake of water and mineral
nutrients.

The only way plant breeders can produce higher yielding
varieties is by producing genotypes possessing combinations of
physiological processes resulting in plants that are structurally

51

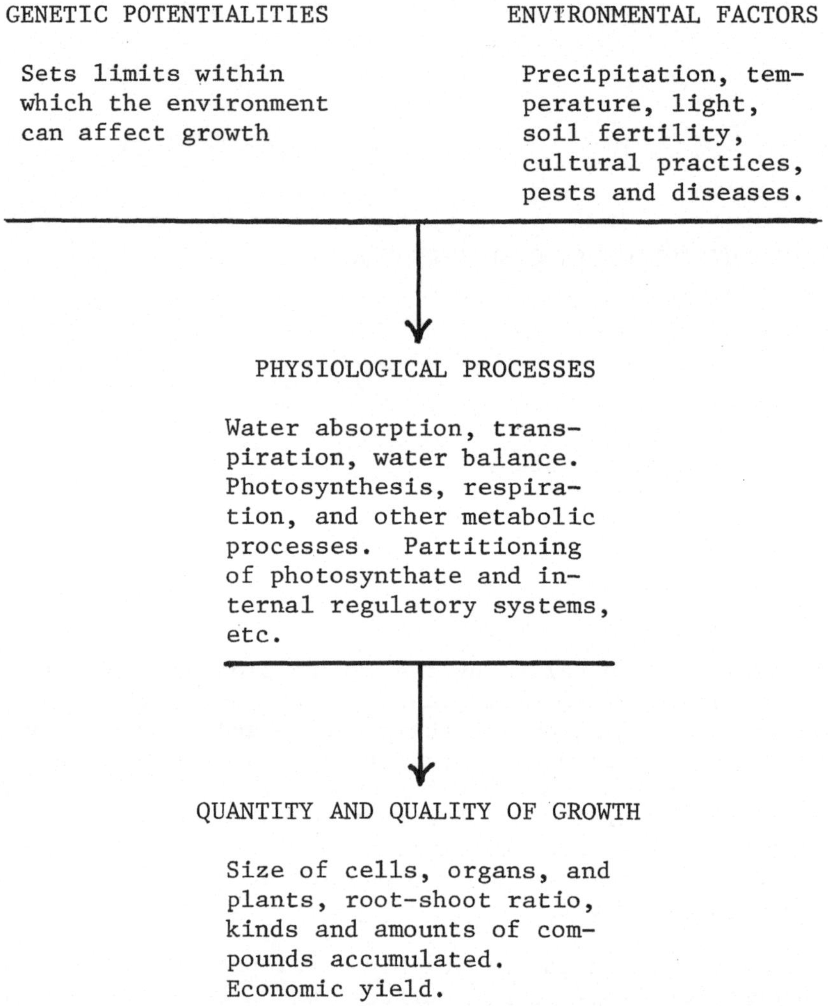

GENETIC POTENTIALITIES   ENVIRONMENTAL FACTORS

   Sets limits within   Precipitation, tem-
   which the environment   perature, light,
   can affect growth   soil fertility,
                                            cultural practices,
                                          pests and diseases.

PHYSIOLOGICAL PROCESSES

   Water absorption, transpiration, water balance.
   Photosynthesis, respiration, and other metabolic
   processes. Partitioning
   of photosynthate and internal regulatory systems,
   etc.

QUANTITY AND QUALITY OF GROWTH

   Size of cells, organs, and
   plants, root-shoot ratio,
   kinds and amounts of compounds accumulated.
   Economic yield.

Figure 1. Genetic potentialities and environmental factors control
           yield through their effects on the physiological
           processes and conditions of a plant.

and/or physiologically more efficient in a particular environment.

    Thus it seems that plant physiology should occupy a central
position in research on crop improvement and the work of agronomists, soil scientists, and plant breeders should be directed
toward development of the most efficient combination of
physiological processes and the most favorable environment
possible for the operation of those processes. Plant physiologists
have accumulated an enormous amount of information about

physiological processes. Some of this has been very important to agriculture, especially in such fields as mineral nutrition, water relations, growth regulators, and herbicides. However, plant physiology has contributed less to agriculture than it could or should, especially in view of its central position shown in Figure 1.

LIMITATIONS ON THE CONTRIBUTIONS OF PLANT PHYSIOLOGY

There are several reasons why plant physiology has contributed less than it should to agriculture. One is that plant physiologists usually are more concerned with obtaining information about physiological processes than with solving plant production problems. They give more attention to the mechanisms of processes such as photosynthesis, respiration, and translocation, than to their relationship with plant growth and yield. Although a few physiologists have worked on the physiology of crop yield since early in the century (Evans, 1975, Chap 1; Watson, 1952), it is only recently that a significant number of plant physiologists have become interested in applying information concerning plant processes to agricultural problems.

The application of basic plant physiology to agriculture probably was delayed by the reductionist philosophy which dominated biology during the 1950s and 1960s and concentrated attention on cellular and molecular biology to the neglect of whole organism biology. This resulted in a generation of physiologists who are relatively ignorant of whole organism physiology. Although molecular and cellular physiology have made important contributions to biological science they have thus far made little or no direct contribution toward increasing crop yields. This is because yield usually is controlled by processes which can be identified at the level of the whole plant such as cell enlargement, leaf expansion, root growth, stomatal opening, and partitioning of photosynthate. It can be argued that all explanations of physiological processes must be reduced ultimately to the molecular level. However, the importance of most environmental stresses with respect to crop yield can be evaluated without an understanding of how they operate at the molecular level.

Finally, it must be admitted that much potentially useful physiological information never is used in agriculture because of the communication gap between laboratory and field scientists. The exchange of information between these two groups usually is left to chance, hence field scientists never learn about useful laboratory research, while laboratory scientists often fail to hear about interesting and important field problems. This communications gap is sometimes aggravated by bureaucratic

specialization and psychological problems. Departmental organization often keeps the two groups of scientists apart, and there also
is an unfortunate tendency for field and laboratory workers to
belittle each other's work. For example, laboratory scientists
often fail to appreciate the importance of problem-oriented field
research and field investigators regard laboratory research as
impractical. Perhaps we need more scientific generalists who can
act as intermediaries between laboratory and field scientists.

PHOTOSYNTHESIS AND YIELD

Experience with the process of photosynthesis illustrates
the difficulty in the conventional approach to applying basic
physiological research to increasing crop yields. The fact that
dry matter production clearly is dependent on photosynthesis led
to the seemingly logical assumption that if the rate of net
photosynthesis could be increased yield also would be increased.
This caused both individual scientists and advisory committees to
urge the U.S. Department of Agriculture and other agricultural
research agencies to conduct more basic research on photosynthesis.
Wade (1973) summarized this criticism and its effects can be seen
in the competitive grants program of the U.S. Department of
Agriculture.

This pressure continues in spite of the fact that English
investigators concluded over 30 years ago that yield rarely is
limited by the potential rate of photosynthesis per unit of leaf
surface. Watson (1952) summarized this research and concluded
that there is little likelihood of materially increasing the rate
of photosynthesis and the best way to increase yield is by
increasing the leaf area index. Loomis, et al., Moss, and Wallace,
et al., in Burris and Black (1975) all point out the difficulty
of relating yield to photosynthesis. Nevertheless, during recent
years there has been extensive research along two lines, search
for plants with high rates of photosynthesis per unit of leaf
surface and search for more efficient mechanisms of photosynthesis.

Unfortunately, this research has shown that yield and photosynthesis are often poorly correlated, both in field crops (Evans,
1975) and in forest trees (Carter, 1972; Helms, 1976). A variety
of shading and defoliation experiments indicate that plants
ranging from apples (Maggs, 1963) and maize (Duncan and Hesketh,
1968) to soybeans (Lenz and Williams, 1973) and wheat (Evans,
1975) have more photosynthetic capacity than is generally used.
While discussing the relationship between photosynthesis and
yield Evans (1975, p. 334) stated "there is little evidence of
any positive relationship between them, nor any instance where
selection for a greater rate of photosynthesis has led to an

increase in yield."

Much effort also has been expended on study of the photosynthetic machinery with the hope of increasing yields. One line of approach was based on the belief that the C4 carbon pathway is more efficient than the C3 pathway and might be transferred to C3 plants such as soybeans by genetic engineering. This transfer seems unlikely because too many processes and too many genes are involved. Furthermore, although the C4 pathway may have some advantages at the level of individual leaves in hot, dry environments it seems to confer little or no advantage at the level of plant stands (Gifford, 1974; Baskin and Baskin, 1978). Another approach involves study of the possibility of conserving carbohydrates by finding a genetic or chemical method of reducing photorespiration (Zelitch, 1975, 1979; Oliver and Zelitch, 1977). However, it seems doubtful if photorespiration can be inhibited without decreasing photosynthesis (Servaites and Ogren, 1977) or disturbing important internal protective mechanisms (Powles and Osmond, 1978). These and other proposals for increasing the efficiency of photosynthesis were discussed in more detail by Zelitch (1975, 1979), Bassham (1977), and Radmer and Kok (1977).

The information obtained from this research is important scientifically, but it has thus far contributed little or nothing toward increasing crop yields. This is partly because the photosynthetic potential of crop plants, expressed as the rate of CO_2 fixed per unit of leaf surface, is seldom limiting, and partly because other processes which follow photosynthesis, such as dark respiration, translocation, and the partitioning of photosynthate to various organs often are as important as photosynthesis itself (Evans, 1975, Chap. 11). Furthermore, under field conditions the rate of photosynthesis is limited more often by environmental stress which causes stomatal closure than by events at the level of chloroplasts. Also, it appears that growth often is more dependent on rapid expansion of leaf area than on the rate of photosynthesis per unit of leaf area (Watson, 1952; Duncan and Hesketh, 1968; Patterson and Flint, 1979). Perhaps plant physiologists interested in increasing crop yield should give more attention to the factors limiting leaf expansion.

YIELD USUALLY IS LIMITED BY ENVIRONMENTAL FACTORS

At the level of the farmer, in the absence of mineral deficiencies, diseases, and insect pests, crop yield is limited chiefly by unfavorable weather. No matter how good the variety or the cultural practices, in the end yield depends chiefly on favorable weather. Thus it seems that the logical place to start research on crop yields is by determining what environmental factors are most limiting for the particular area and for the

particular crop under study. The two most commonly limiting
environmental variables are temperature and water and these will
be discussed in detail.

Temperature Perturbations

There has been considerable discussion concerning the
probable effects on plant growth and crop yield of a small
increase or decrease in global temperature. Small changes in
global temperature may be important in the long run. However,
short term exposure to periods of abnormally high or low temperature at various stages of growth is of much more immediate
importance than small global trends because short term perturbations can have disproportionately large effects on growth. A
week of cold weather just after cotton is planted or tobacco is
transplanted to the field, or a week of hot, dry weather at the
time when pollination is occurring in corn will have much more
effect on yield than a deviation of one or two degrees above or
below average during the entire growing season. For example,
three days exposure to temperatures 3 or 4° below normal field
temperatures reduces growth of cotton more than the growth of the
competing weeds, velvet leaf and spurred anoda (Patterson and
Flint, 1979). Exposure to a week or ten days of temperature
only 4° lower than normal just after transplanting increases the
ratio of length to width of tobacco leaves (Raper, 1972). There
also are reports that the temperature during seed formation has
significant effects on seed development, germination, and seedling
growth. Examples are research of Akpan and Bean (1977) on pasture
grasses; Powell and Huffman (1978) on sorghum; and Thomas and
Raper (1975) on tobacco. Injury from exposure to abnormal
temperatures has been attributed to damage to the photosynthetic
apparatus (McWilliam and Ferrar, 1974). However, some of the
injury may be related to development of water stress caused by
reduced water absorption at low temperatures. This has been
suggested for cotton (Arndt, 1937; St. John and Christiansen,
1976), beans (Crookston, et al., 1974) and soybeans (Taylor and
Rowley, 1971).

If more were known about the manner in which these temperature perturbations affect physiological processes it might be
possible to develop varieties with more tolerance. Here is where
plant physiologists can and are making important contributions by
investigating the causes of injury.

Water Stress

In the field, photosynthesis and other essential physiological
processes probably are reduced more often by water stress than by

all other environmental factors combined. Although water stress
affects every aspect of plant physiology and growth, a major part
of the reduction in growth and yield is related to reduction in
photosynthesis. This reduction is chiefly caused in three ways:

    1. By reducing leaf size and photosynthetic surface.
    2. By closure of stomata, reducing the supply of CO_2
       reaching the chloroplasts.
    3. By damage to chloroplast structure.

The extensive literature on the effects of water stress on
photosynthesis and other physiological processes was reviewed by
Begg and Turner (1976), Boyer (1976), and Hsiao (1973). It is
obvious that from the physiological viewpoint more research is
needed to explain how moderate water stress inhibits various
plant processes at the cellular and molecular level. We also
need to know more about dehydration tolerance, about stomatal and
cuticular control of water loss and CO_2 exchange, osmotic adjust-
ment, and about such whole plant characteristics as depth and
extent of rooting and root permeability. However, injury to crop
plants by water stress is more often caused by limitation of leaf
area expansion or stomatal closure than by events at the
molecular level, such as electron transfer.

HOW TO INCREASE THE CONTRIBUTIONS OF PLANT PHYSIOLOGY

In the past plant physiology has often provided explanations
for subnormal crop yields and sometimes has indicated remedies.
In the future it ought to be used to predict what physiological
processes are likely to be limiting to plant yields in various
soils and climatic conditions. Such information would aid plant
breeders to produce varieties containing the optimum combination
of physiological and morphological characteristics required to
tolerate the environmental stresses of particular environments.
Evans (1975, Chap. 1) suggests that emphasis has been on morphological characters such as short stiff stems and canopy geometry.
However, in the future there should be more emphasis on inter-
relationships among processes such as photosynthesis, translocation, and sink size and strength which affect the partitioning
of photosynthate; also synthetic processes which affect chemical
composition and quality of yield.

As the physiological processes which control yield are
greatly affected by the environment, the most effective contributions to plant production are likely to be made by research in
environmental and stress physiology. This requires a different
approach from that traditionally used in plant physiology, where
the primary emphasis usually is placed on plant processes. The
first step is identification of the environmental stresses which

are limiting yield in a particular situation. This should be
followed by controlled environment and laboratory research to
identify the physiological processes through which the environmental stresses are inhibiting growth and crop yield. Then the
plant breeder can produce the particular combination of physiological characteristics which will enable plants to tolerate a
particular stress with minimum injury.

   This approach requires the organization of interdisciplinary
teams to solve particular problems of plant production. These
teams ordinarily should consist of agronomists, soil scientists,
agricultural meteorologists, plant physiologists, and plant
breeders. The agronomists should be able to identify situations
where crop yield is reduced, the soil scientists and meteorologists
will determine what environmental stresses are limiting yield,
and the frequency and severity of these stresses. The plant
physiologists will determine which physiological processes are
being affected and at what stage of growth the most damage occurs.
For example drought during the reproductive stage usually does
more damage than earlier or later in the life cycle. Once the
problem is identified in terms of physiological processes plant
breeders can search more effectively for the characteristics
which will enable plants to tolerate the limiting condition.

   For example, such a characteristic as drought tolerance is
too broad and general to be used as a basis for plant breeding
and selection. Specifically, do we need deep root systems, more
root branching, more permeable roots, responsive stomata and
thick cuticle, dehydration tolerance, or what? If it is early
maturity, then low temperature tolerance in the seedling stage
may be the key characteristic.

   It is difficult to organize this kind of problem-oriented,
interdisciplinary team research because by disposition and
training most scientists are individualists who prefer to work
on problems of their own choice. Part of this individualistic
tendency results from the manner in which we train Ph.D.s. Most
of them are allowed to pursue their graduate training with little
or no interaction with the research of other graduate students.
Perhaps we should introduce more graduate students to team
research projects early in their career so they can see its
advantages and become more willing to participate in cooperative
research. Russell (1978) recently published an interesting
discussion of the problems involved in organizing an interdisciplinary, problem-oriented research program, and pointed out
that it can be successful and satisfying after the psychological
barriers are surmounted.

We also need to broaden the mental perspective and versatility of graduate students, especially those in plant physiology, by exposing them to more field and whole plant problems. Usually their experience is limited to laboratory research on a single process such as photosynthesis, or in a limited area such as membrane physiology, electron transfer, a particular enzyme system, or the stomatal mechanism. However, they often neglect the relationship of their research to the larger problems of plant growth. Such concentration in a narrow field may be conducive to establishing a reputation in research, but it will never solve the problems of plant production. These problems are so broad and so complex that their solution requires multi-disciplinary research by teams of versatile, broadly trained scientists.

GENERAL DISCUSSION

These remarks are made in the context of a symposium on the relationship of basic research to crop improvement in less developed countries. However, most of them apply equally well in fully developed countries. When specific problems are awaiting solution we need to carry on the kind of research most likely to solve them. We cannot assume that merely increasing the amount of basic research will serendipitously provide the information required to increase crop production. Rather we must identify the components of the problem, assemble the kinds of scientists who are most likely to contribute to its solution, and put them to work in a problem-oriented program to find a solution.

This problem-oriented approach may seem to depreciate the importance of basic research, but that is not my intention. We need basic research because it provides the foundation of information essential for problem solving. Furthermore, most important advances in science are made by creative, imaginative individuals who are allowed to follow their own inclinations. We always need more of that kind of research. However, basic research ought to be supported in its own right as an important intellectual activity, and not under the pretense that it will increase food production, cure cancer, or increase the gross national product. Some may, but much will not.

When money, scientific manpower, and time are limiting priorities become necessary and solution of crop production problems becomes more important than mere increase in scientific knowledge. However, in the long run basic research probably will benefit because scientists gifted with imagination will find that during the solution of applied problems they will identify new areas of plant physiology which deserve investigation. For example the practical importance of photoperiod in ornamental

horticulture and floriculture stimulated research in photo-
physiology and the discovery of phytochrome.

What we really need is productive research on important
problems and whether it is termed "basic" or "applied" is of
little importance. The essential thing is to determine what needs
to be investigated and then to find an effective method to work
on it. I believe that a multidisciplinary approach which starts
in the field and works back to the laboratory is the most
effective way to deal with plant production problems in any
country, whether it is developing or already developed.

SUMMARY

Physiological processes occupy a central position in plant
growth because they constitute the machinery through which the
genotype and the environment affect the quantity and quality of
growth. Nevertheless, plant physiology has contributed less to
crop production than it should. This is partly because it is
process oriented rather than plant growth oriented, partly
because much good physiological research has no direct bearing on
crop yield, and partly because a communications gap often exists
between plant physiologists and crop scientists.

For example, the scientifically important research on the
photosynthetic machinery has thus far contributed little or
nothing toward increasing crop yield. Most crop plants already
have a high potential rate of photosynthesis, but the potential is
seldom attained because of environmental limitations.

In general, crop yields are limited more often by unfavorable
weather than by lack of capacity in physiological processes. Thus
to improve yield we need a better understanding of how environmental factors such as water and temperature stress affect the
physiological processes which control growth and yield. Such
information will provide specific objectives for plant breeding
programs much more useful than such general terms as drought or
heat tolerance.

To obtain this information requires an approach quite
different from that traditionally used in plant physiology.
Research must start in the field where the limiting environmental
factors can be identified. Then controlled-environment and
laboratory research can determine what physiological processes are
being inhibited and provide plant breeders with specific objectives.

This approach requires interdisciplinary team research
involving agronomists, soil scientists, meteorologists,

physiologists, and plant breeders willing to cooperate in solving specific problems. Such research is difficult to organize but it can be more effective in solving plant production problems than traditional approaches. Furthermore, in the long run it will inevitably generate much interesting basic research.

LITERATURE CITED

Akpan, E. E. J., and Bean, E. W., 1977, The effects of temperature upon seed development in three species of forage grasses, Ann. Bot., 41:689-695.

Arndt, C. H., 1937, Water absorption in the cotton plant as affected by soil and water temperatures, Plant Physiol., 12: 703-720.

Baskin, J. M., and Baskin, C. C., 1978, A discussion of the growth and competitive ability of C3 and C4 plants, Castanea, 43:71-76.

Bassham, J. A., 1977, Increasing crop production through more controlled photosynthesis, Science, 197:630-638.

Begg, J. E., and Turner, N. C., 1976, Crop water deficits, Adv. Agron., 28:161-217.

Boyer, J. S., 1976, Water deficits and photosynthesis, in: "Water Deficits and Plant Growth," Vol. 4, pp. 154-190, T. T. Kozlowski, ed., Academic Press, New York.

Burris, R. H., and Black, C. C., eds., 1975, "CO2 Metabolism and Plant Productivity," University Park Press, Baltimore.

Carter, M. C., 1972, Net photosynthesis in trees, in: "Net Carbon Assimilation in Higher Plants," pp. 54-74, C. Black, ed., Am. Soc. Plant Physiol., Atlanta, Georgia.

Crookston, R.K., O'Toole, J., Lee, R., Ozbun, J. L., and Wallace, D. H., 1974, Photosynthetic depression in beans after exposure for one night, Crop Sci., 14:457-464.

Duncan, W. G., and Hesketh, J. D., 1968, Net photosynthetic rates, relative leaf growth rates and leaf numbers of 22 races of maize grown at eight temperatures, Crop Sci., 8:670-674.

Evans, L. T., ed., 1975, "Crop Physiology," Cambridge University Press, London.

Gifford, R. M., 1974, A comparison of potential photosynthesis, productivity and yield of plant species with different photosynthetic metabolism, Aus. J. Plant Physiol., 1:107-117.

Helms, J. A., 1976, Factors influencing net photosynthesis in trees: an ecological viewpoint, in: "Tree Physiology and Yield Improvement," pp. 55-78, M. G. R. Cannell and F. T. Last, eds., Academic Press, London.

Hsiao, T. C., 1973, Plant responses to water stress, Annu. Rev. Plant Physiol., 24:519-570.

Lenz, F., and Williams, C. N., 1973, Effect of fruit removal on net assimilation and gaseous diffusive resistance of soybean leaves, Angew. Botanik, 47:57-63.

Maggs, D. H., 1963, The reduction in growth of apple trees
    brought about by fruiting, J. Hort. Sci., 38:119-128.
McWilliam, J. P., and Ferrar, P. J.,1974, Photosynthetic
    adaptation of higher plants to thermal stress, in:
    "Mechanisms of Regulation of Plant Growth," pp. 467-476,
    R. L. Bieleski, A. R. Ferguson, and M. M. Cresswell, eds.,
    Roy. Soc. New Zealand, Wellington.
Oliver, D. J., and Zelitch, I., 1977, Increasing photosynthesis
    by inhibiting photorespiration with glyoxylate, Science, 196:
    1450-1451.
Patterson, D. T., and Flint, E. P., 1979, Effects of chilling on
    cotton (Gossypium hirsutum), velvetleaf (Abutilon theophrasti),
    and spurred anoda (Anoda cristata), Weed Science, 27:in press.
Powell, R. D., and Huffman, K. W., 1978, The development of the
    small seed syndrome in sorghum in relation to environmental
    conditions, Plant Physiol. Supplement to Vol. 61, p. 5.
Powles, S. B., and Osmond, C. B., 1978, Inhibition of the capacity
    and efficiency of photosynthesis in bean leaflets illuminated
    in a CO2-free atmosphere at low oxygen: a possible role for
    photorespiration, Aus. J. P. Physiol., 5: 619-629.
Radmer, R., and Kok, B., 1977, Photosynthesis: limited yields,
    unlimited dreams, BioScience, 27:599-605.
Raper, C. D., Jr., 1972, Temperatures in early post-transplant
    growth: alteration of leaf shape in field environments,
    Tobacco Sci., 17: 14-16.
Russell, R. S., 1978, Co-ordination and redeployment in research,
    1977 Ann. Rep. Letcombe Laboratory, Wantage, England.
Servaites, J. C., and Ogren, W. L., 1977, Chemical inhibition of
    the glycolate pathway in soybean leaf cells, Plant Physiol.,
    60:461-466.
St. John, J. B., and Christiansen, M. N., 1976, Inhibition of
    linolenic acid synthesis and modification of chilling
    resistance in cotton seedlings, Plant Physiol., 57:257-259.
Taylor, A. O., and Rowley, J. A., 1971, Plants under climatic
    stress. I. Low temperature, high light effects on photo-
    synthesis, Plant Physiol., 47:713-718.
Thomas, J. F., and Raper, C. D., Jr., 1975, Differences in
    progeny of tobacco due to treatment of the mother plant,
    Tobacco Sci., 19:37-41.
Wade, N., 1973, Agriculture: critics find basic research stunted
    and wilting, Science, 180:390-393.
Watson, D. J., 1952, The physiological basis of variation in yield,
    Adv. in Agron., 4:101-145.
Zelitch, I., 1975, Improving the efficiency of photosynthesis,
    Science, 188:626-633.
Zelitch, I., 1979, Photosynthesis and plant productivity, Chem. &
    Eng. News, 57(6):28-32, 37-42, 46-48.

BLUE ROSES AND BLACK TULIPS:

IS THE NEW PLANT GENETICS ONLY ORNAMENTAL?

Peter S. Carlson

Department of Crop and Soil Sciences
Michigan State University
East Lansing, MI   48824

INTRODUCTION

There have been, and I expect will continue to be, discussions concerning the relevance of the newly-emerging techniques in plant genetics (viz. in vitro cell culture, somatic cell genetics, recombinant DNA, and the application of the experimental systems of molecular biology) for crop improvement. From a number of recent reports and recommendations has come the expectation and even the anticipation that newly-emerging analytic and genetic biological techniques will contribute to the production of improved varieties of crop species (Brown, et al., 1975). Can the experimental approach and reductionistic assumptions of molecular biology be utilized for the solution of problems in agricultural plant biology and crop improvement? Will a correlation of in vitro events with the responses of crop plants in the field allow a better understanding and, more importantly, allow manipulation of biological processes underlying crop productivity?

There are several responses to these questions, each of which have been expressed in one form or another during the numerous recent debates concerning the potential of increasing agricultural productivity. The first response points out that our current levels of crop productivity were achieved in the absence of a direct knowledge of molecular mechanisms, and that there is no reason to expect that this kind of new knowledge will enhance productivity. Only a few agronomically useful varieties of field crops have resulted from the use of "molecular" techniques; hence these techniques are largely irrelevant to applied agricultural plant biology. I suspect that this response might be termed the reactionary right.

63

A second response, which is the direct opposite of the first, asserts that only by a complete molecular analysis of the processes underlying crop productivity is there any hope of manipulating the individual components of crop yield in a rational way. Plant breeders have nearly exhausted the wild germplasm in many crop species; hence, the new genetics also holds out the promise of otherwise unavailable genetic variability. This response might well be termed the radical left. A third response suggests that a molecular analysis will be of importance in manipulating some biological processes, but will not be a panacea for all the problems of agricultural biology. This is the position of the pragmatic moderate...the one who falls into the catch-all category of realist. Since I fall into the third group, it seems appropriate to define in which situations I think molecular analysis may prove important and in which situations it may be irrelevant. The following pages will be devoted to an attempt to bring some discrimination to my response.

This document is intended to explain these new genetic technologies and to explicate their potential impact to those whose expertise lies outside the biological sciences. The direction and extent of scientific effort and the resulting availability of new technologies is conditioned in large measure by economic, political and social forces. Science does not operate in a vacuum. Further, communication among disciplines is necessary for any integrated, informed approach to change.

The focus of the symposium is an examination of the role of basic research in altering the life situation of small-holding, primarily marginal farmers of the third world. Plant genetics, in actual fact, will probably play only a minor role in changing the condition of such individuals when compared to political and economic forces. However, the role of plant genetics appears real and not inconsequential: improved seed are an often cited need for these farmers. Improved seed probably holds as much potential for helping small farmers as any other single agronomic input. What types of genetic alterations are appropriate for the needs of third-world agriculture? Certainly the agronomic situation of the small, marginal farmer is different from that of the large commercial farmer common in the western world. It is impossible to transfer western farming technology if there is not the ready availability of chemical and physical means to manipulate the environment. Additionally, the economic (e.g., market price, access to credit) and political (e.g., land tenure arrangements) constraints are beyond immediate control.

Small farmers have responded with a diversity of solutions to their individual situations. It appears to me as if there is one central theme in all of the various viable solutions: to develop a defensive stance. In agronomic terms this may mean a diversity

of crops for home consumption and for the market. In biological
terms this implies genetic variability. Some risk is acceptable,
but not too much. Some harvest is preferable to no harvest, and
when there is little control over the environment, the genetic
composition of the crop planted becomes an important consideration.
Highly inbred and uniform varieties or hybrid varieties are often
inappropriate for these demands. Genetic variability such as that
found in land races is perhaps more adaptive to the situation.
This defensive stance may alter the plant breeder's strategy: increasing yield, while continuing to be an essential breeding
objective, becomes of secondary importance when compared with survival and adequate levels of production under the entire range of
possible environmental (physical, chemical, and biological) conditions. Hence, a major concern for crop improvement must be to
maintain and increase the range of genetic variability which permits a crop species to survive and be fairly productive under
widely different conditions. The production and analysis of genetic
variability and its incorporation into adapted varieties is an
appropriate goal to aid these farmers.

     The inherent strength of conventional plant breeding techniques
and new genetic techniques for the small farmer is that they are
potentially scale neutral, politically neutral, relatively inexpensive for the farmer, and can involve few ties with commercial networks. These qualities make plant improvement programs and their
results accessible to small farmers and acceptable to governmental
agencies and international developmental efforts. While the crop
improvements provided by the "green revolution" have not always
been of this nature, there is no reason why the techniques of breeding and genetics cannot accomplish these goals.

SOME OF THE COMPONENTS OF THE NEW GENETICS

     Recent advances in molecular and cellular biology have provided
methods of genetic manipulation which should be applicable to the
agronomic improvement of plant species (Carlson and Polacco, 1975;
Bottino, 1975; Bhojwani et al., 1977; Kleinhofs and Behki, 1977;
Thomas et al., 1979). Despite the rapid expansion in our knowledge of
basic genetic and biochemical mechanisms in lower organisms, such
as bacteria and fungi, this knowledge has had no direct impact on
plant improvement. Such a lack of impact may be ascribed in large
part to conceptual and experimental differences between the disciplines of molecular biology and plant breeding. Molecular biology
is comprised of two basic elements: the reductionist approach of
basic science derived from physics and chemistry and a powerful
set of analytical experimental tools which arose from microbial
biochemistry, physiology and genetics. One example of the utility
of this approach was the use of defined genetic variants combined
with precise biochemical methods to elucidate various metabolic

pathways and the mechanisms by which they are regulated in a variety
of microorganisms. The majority of our current ideas concerning
biology have been derived from or shaped by this discipline. In
contrast, plant improvement as currently practiced has of necessity
a more holistic approach. Plant breeders have to operate within
difficult constraints. They have little choice in either their
experimental materials or the problems which confront them.

Crop species were domesticated during the Stone Age because
of their characteristics which allowed them to be rapidly grown and
harvested in abundance; not because of ready manipulation in scientific experiments. Plant breeders must respond to an entire range
of possible pressing problems ranging from biochemistry to pathology
to the needs of agricultural engineering and food processing.
Additionally, strong correlations have yet to be established
between crop yield and any of the individual physiological or biochemical processes that contribute to the final harvested product.
The experimental and technological possibilities of plant breeding
and the inherent constraints of the plant system are very different
from those required for a direct molecular biological analysis.

Two different approaches which attempt to extend the experimental methods and techniques of molecular biology from microbial
investigations to applications for crop improvement are presently
being developed: one involves cellular manipulations and the other
involves DNA manipulations. Cellular manipulations hold the potential for developing new experimental systems utilizing crop
species which are suitable for more refined analytical techniques.
Using single somatic cells of plants, it is possible to achieve
mutant production, analysis and hybridizations not possible using
whole plants. The primary elements in an ideal and defined cellular system include: a) ready availability of haploid material;
b) easy production and maintenance of suspension cultures; c) rapid
production, manipulation and culture of protoplasts and d) routine
ability to regenerate entire plants from cells cultured in vitro.

Direct manipulation of DNA (i.e. recombinant DNA techniques)
has great potential for use in the identification and modification
of defined genetic elements, although much work needs to be done
before such sophisticated techniques can be applied to specific
problems of plant biology. The primary elements in a defined DNA
manipulation system would include: a) routine techniques for
cloning plant DNA; b) microbial systems which permit the expression
of plant DNA; and c) definition of cloning vehicles for or of
genetic transformation in higher plants.

Although many tools of the molecular biologist are now
becoming available to the plant biologist, several limitations
prevent their direct application to specific breeding problems

The first problem with these approaches is a technical one:
regeneration of whole plants from single cells is essential for
application of the technology of in vitro genetic manipulation of
higher plants. However, this step has only recently been accomplished with a few major food crops; it is as yet just "potentially
possible" with most agronomic species. Plant breeders must be
able to incorporate the results of any genetic modification into
their breeding programs via sexual techniques. Gamates are produced
by mature plants and not directly from cells cultured in vitro.
And of course the actual agronomic impact of any genetic modification produced at the level of the cell must be assessed at the
whole plant level.

The second problem results from the developmental biology of
agronomic characters. Many agronomic traits are tissue-specific;
their expression is found in only one or a few tissues within the
plant and is often not found in cells cultured in vitro. Cell cultures represent a distinct type of biological situation which is
not found in the whole plant. These cells are not biochemically
or physiologically identical to cells found in roots, leaves, shoots
or flowers. If a particular trait is not expressed in culture,
there is no reason to expect that the trait can be altered and identified via in vitro methods. Hence, if a particular characteristic
is found primarily in the leaves or shoots or roots of a species,
in vitro cultures probably do not provide an adequate experimental
situation to investigate that characteristic.

The third problem involves the genetics of agricultural traits.
Cell culture systems and DNA manipulations allow modification of
only single gene-controlled traits. Most agronomic traits, as they
are now defined, are polygenic in inheritance. A polygene controls
small, additive, stepwise alterations which are impossible to recog-
nize individually. Characters such as yield, response to stress
conditions like drought or salt, and plant architecture are each
determined by a large number of genes. Hence these characters cannot be analyzed or altered with these experimental systems. Currently, genetic modification of crop plants using cellular or DNA
manipulations should prove appropriate in cases where the desired
alteration involves single gene traits which are not tissue-specific
and for which there are adequate, chemical-selective techniques.
The technology involved in these new approaches will almost certainly be improved to overcome the present limitations. However,
at present, single gene traits which are not tissue-specific are
rare, as are appropriate selective systems with which to recover
genetic variants in these characters.

Despite these current limitations, there are strong reasons to
expect that these techniques will be of importance in plant improvement programs. First, the majority of these cellular and DNA techniques have been developed over the last ten years; productive

experimental effort is currently focused on the problems posed by regeneration, developmental biology and quantitative genetic characters. Second, and perhaps more important, it is not necessary to have completely defined cellular or DNA manipulation systems for these techniques to be of use in plant improvement.

The current limitations of these new technologies are severe when viewed from the perspective of third world farmers. The great majority of experimentation with these new techniques has been done with model systems, plants with little or no agronomic importance, or with the major agronomic species of the "developed" world. There is much less and, in many cases, no information concerning the staple crops or the cash crops of importance to small third world farmers. In a very real agronomic sense, if it is not corn, wheat, rice or potatoes, we do not know about its genetics and we have not defined methods for its genetic modification. In this regard, the new technologies do not substantially differ from many standard breeding programs. There is a real need for genetic and breeding information in crops other than those common in western agriculture.

I feel that until the use of these new techniques has resulted in the production of agronomically useful crop varieties, the burden of proof for their potential utility resides with basic scientists. Plant breeders currently shoulder many responsibilities: they cannot be expected to utilize essentially unproven techniques. However, basic scientists need and should be able to solicit the support and cooperation of plant breeders.

COMPLETELY DEFINED CELLULAR SYSTEMS ARE NOT REQUIRED FOR USE IN PLANT IMPROVEMENT.

Many novel genetic manipulations of importance for plant breeding can be developed using these new methods even though completely defined cellular systems or methods for DNA modification are not yet available for crop species. The important components in defining the use of these methods are cooperation between plant breeders and geneticists, their use of a common vocabulary, and the willingness to fully employ currently available techniques. Unfortunately, such cooperation has rarely occurred. Breeders usually cannot or do not make an effort to learn what can be currently accomplished, and consequently they fail to consider how these manipulations may fit into their crop improvement programs. Cell and molecular biologists often do not talk with or attempt to understand the needs of breeders. Additionally, they assume that they must have all of the components of their experimental system defined in detail before they are willing to consider a strategy for plant improvement. This is an unfortunate situation. Instead of producing ever-increasing numbers of agronomically irrelevant somatic hybrids and cellular variants,

basic scientists must use these model systems to develop experimental systems with relevance to plant improvement.

There are a small but growing number of collaborative efforts
between plant cell biologists and breeders. Several examples will
be described which have been successful or appear near success. In
these examples, routinely available cellular techniques were used
directly or only slightly modified to fit into the needs of plant
improvement programs. Several of these efforts have resulted in
genetic manipulations which cannot be achieved utilizing only conventional breeding methods.

The most successful application of _in vitro_ techniques has
been the development of rapid-cloning systems by utilization of
meristem and shoot tip culture and subsequent rapid multiplication.
These techniques have resulted in a marked change in the floriculture industry. Rapid multiplication of apical meristems permits
the routine production of a large number of genetically uniform
plants. These techniques are currently available for a number of
horticulturally important species and are used extensively in the
production of some cut flowers and potted plants. The techniques
of meristem culture have been employed with some species, most
notably with potato, in the production of virus-free crop plants.
The technique of rapid multiplication of genetically identical individuals has also been developed with several tree species and pro-
mises to be a central technique in the genetic improvement of forest
trees.

Another example of the successful utilization of these techniques for crop improvement comes from work with rice. New varieties of this crop species have been developed by the direct application of anther culture which permit the rapid production of haploid
plants. Identification of superior haploid individuals produced
from highly heterozygous diploid individuals, combined with subsequent chromosome doubling, has resulted in several genetically
homozygous varieties. Improved rice varieties produced via these
manipulations have been released in China. As a result, new
varieties can be developed in four or five years, and the time
required for the breeding cycle is much reduced. Similar novel
approaches to breeding wheat and tobacco via haploid selections
have proven successful (Han et al., 1978).

Several experimental situations have been described which
allow the direct recovery of useful genetic variability from either
cells or protoplasts cultured _in vitro_. In sugar cane and potato,
clones more resistant to disease have been identified after regeneration of cells from culture. These findings are presumed to be
the result of a high rate of genetic variability which naturally
occurs in cells cultured _in vitro_ (Heinz et al., 1977).

The following two examples of the utility of in vitro techniques for crop improvement are taken from ongoing programs at Michigan State University. Hordeum vulgare, common barley (2N=14), has a narrow germplasm base as compared with other cereal grains. Improved varieties of barley would result if it were possible to incorporate genetic resistance to cereal leaf beetle and increased tolerance to freezing, drought and high salt. All of these desirable characteristics are present in a second species of Hordeum, H. jubatum (2N=28). A number of F_1 hybrids have been produced between these two species, and the karyotype of the resulting hybrid consists of 21 chromosomes. The hybrids are completely sterile and have resisted all attempts to double the chromosomes in the entire plant. Cell culture efforts were first utilized in an attempt to double the chromosomes of this hybrid to restore its fertility. It is currently possible to regenerate entire barley plants from callus derived from immature ovaries. Callus was produced from the immature ovaries of the F_1 hybrid plant, the callus was treated with colchicine, and a number of plants were regenerated. Among the regenerated plants were some with a doubled chromosome number (42 chromosomes). Unfortunately, these plants were also sterile. This situation was more complex than expected in a usual amphidiploid. The next step in our approach was to make use of a usually bothersome characteristic of cells cultured in vitro: a karyotypic instability. Cultured cells undergo rather rapid changes in their chromosomal constitution, producing both diploid and aneuploid alterations.

The original breeding intention was to introgress the interesting characters from H. jubatum via the fertile amphidiploid from the F_1 into H. vulgare by continued backcrossing to the H. vulgare parent. This manipulation is impossible because of the continued sterility of the F_1 and its amphidiploids. However, the same end result can be achieved by using cultured cells and focusing on their characteristic of karyotypic instability. Entire plants can be recovered from the cultured F_1 hybrid callus that contain almost complete haploid genetic components from H. vulgare with only small quantities of genetic information from H. jubatum. That is, we can utilize karyotypic instability in cell culture to eliminate a majority of the H. jubatum component while maintaining the H. vulgare component. Such plants would be recovered when large populations of regenerated plants are screened. And indeed, just such individuals have been recovered. They have seven chromosomes, are morphologically similar to H. vulgare, however, they express several isozymes which are characteristic of H. jubatum. The chromosomes of these plants have been doubled, and near diploid seeds have been produced. Plants resulting from these seeds will be screened for the interesting characteristics of H. jubatum. Lines with the desirable characteristics of H. jubatum will be directly utilized in an ongoing breeding program. Hence, it is possible to achieve

"introgression" in vitro. Karyotypic variability can be utilized
to circumvent the absolute block to progress imposed by hybrid
sterility (Orton, unpublished).

Numerous attempts have been made to recover fertile interspecific hybrids within the Leguminosae, including crosses of Vigna
radiata, mung bean, by either V. umbellata, rice bean, or V. mungo,
black gram. No useful hybrids have been obtained from this work.
The application of in vitro techniques to interspecific hybridization
within the genus Vigna has recently yielded potentially valuable
hybrid material. Supplemental techniques were used to overcome
crossability barriers between V. radiata and V. umbellata: chemical
treatment of parental plants, application of nutrient solutions
to the female parent's stigmatic surface, defoliation after cross-
pollination, the use of mixed pollen, including that of a third
species, and most importantly, in vitro culture of hybrid embryos
and pods.

Excision and in vitro culture of hybrid embryos prior to the
abortion which occurs some 10-16 days post-pollination permitted
recovery of a number of hybrid plants. Embryos of putative hybrid
origin were excised 11-14 days post-pollination and cultured in
vitro. Aseptic transfer of plantlets to sterile soil and their
gradual acclimation to ambient laboratory and greenhouse conditions
yielded a number of vigorous interspecific Vigna hybrids. The recovery of these hybrids in large number is the result of the rather
simple incorporation of refined methods for in vitro manipulation:
the culture of young embryos in these species at a stage before
abortion in the F_1 hybrids.

The hybridity of interspecific F_1 seedlings produced from the
cross of V. radiata x V. umbellata was verified by cytogenetic analysis and by genetic markers. However, these hybrid plants were
both male and female sterile. The ploidy of one such sterile hybrid
plant was doubled with colchicine in vitro to produce an amphidiploid with a degree of fertility. This amphidiploid was backcrossed
to V. radiata to produce allotriploids which were recovered with
the aid of embryo culture techniques. The allotriploid plants grew
vigorously and flowered profusely with a very low level of seed fertility. A number of seed, however, were produced from self-pollina-
tions and backcrosses to V. radiata with these allotriploids.
Samples of this newly derived Vigna material are being sent to the
Asian Vegetable Research and Development Center in Taiwan. There
the material is being screened for the presence of valuable rice
bean characteristics in the mung bean genetic background (Baker
et al., unpublished).

WHAT KIND OF GENETIC VARIABILITY IS AGRONOMICALLY DESIRABLE?

Plant breeders search for and recover genetic combinations which display desirable traits under certain environmental conditions. Their approach involves the production of populations with a broad genetic base, followed by selection at the whole-plant level for recombinants with desirable alterations. Genetic manipulation is practiced without knowing the biochemical basis of the separate components which comprise the character being modified. The focus of breeding efforts is centered on selecting the desirable recombinant types that emerge from any particular cross or segregating population. Currently, the assays of agronomic utility and the subsequent selections are based on observations of whole plant phenotypes.

The complexity of plant biology and of plant productivity is expressed in the genetics of most agriculturally important traits. These traits appear to be controlled by "polygenes" and their transmission is analyzed by quantitative methods. Significant progress in plant breeding could be made in the improvement of breeding techniques if it were possible to establish reliably physiological or biochemical assays at critical points in a number of the individual component processes of agronomic traits. With critical processes individually analyzed and assayed, various genotypes, each demonstrating optimal performance at a different step in a process, could be combined to produce a most productive cultivar.

This type of effective genetic manipulation and breeding of plant productivity requires identification of the relevant metabolic processes and specific rate-limiting steps within these processes. However, despite considerable effort, final yield has not yet been found to be strongly correlated with any distinct biochemical pathway or biochemical characteristic. The problem is that a multitude of biochemical reactions can affect the final productivity. The problem is in identifying which reactions or steps actually do affect productivity under field conditions.

A characterization of heterosis may provide information about the individual processes that limit expression of quantitative or polygenic traits. Heterosis refers to the beneficial effects (increased size and productivity) which are observed in some F_1 hybrids derived from crosses between unrelated parental lines. This classically defined genetic phenomenon is described almost entirely by statistical methods and remains essentially uncharacterized in biochemical or molecular terms. Known genetic and molecular mechanisms can account for the phenomenon of heterosis. However, there is no direct evidence that any of these mechanisms are in fact involved in heterosis.

Several cases have been described where the activity of a specific biochemical enzyme or pathway (e.g., nitrate reductase or mitochondrial oxidative phosphorylation) was higher in the hybrid than in either of the parental lines. However, these experiments are only circumstantial in nature, and no causal relationship has been established between such increases in enzyme activity and the overall increases observed in growth or yield. Despite uncertainty concerning the mechanisms involved, it is clear that constraints in one or more of the genetically controlled biochemical character- istics that limit yield in the parents have been relieved in a superior hybrid. Hence, such hybrids offer an opportunity to iden- tify biochemical traits that are of direct importance to the rate- limiting steps contributing to these traits. Biochemical or physio- logical steps which were rate-limiting in the inbred parental lines have been relieved in the hybrid; other biochemical limitations determine the growth rate of the heterotic hybrid.

What is needed is an experimental system dealing with a specific character that satisfies both the classical definition of a heterotic effect and the requirements for effective manipulation and analysis of the discrete genetic and biochemical components involved. One approach would be to use specific stress conditions to modulate the magnitude of the hybrid advantage. For example, if the superior per- formance of a hybrid genotype is due in part to more efficient con- version of a particular input (i.e., light, CO_2, nitrate, phosphate, etc.) to a useful product, than 1) the heterotic advantage should diminish in the total absence of that input; 2) the hybrid advantage should be magnified under limiting conditions where growth is di- rectly related to the level of that specific input (i.e., stress conditions); and 3) the relative response of the hybrid and parental genotypes should be unaffected if the heterotic effect is not re- lated to the stress conditions. It should, therefore, be possible to take a hybrid clearly showing heterosis under field conditions, raise it under controlled specified stress conditions in the labor- atory, and thereby identify parameters which alter the relative per- formance of the hybrid and parental genotypes. Using this approach, it should be possible to pinpoint specific biochemical pathways that contribute to hybrid vigor. Extended study can focus on indi- vidual steps within those pathways, elucidate the biochemical mech- anisms responsible for heterosis, and identify the rate-limiting steps underlying final productivity under stress conditions.

This experimental approach was tested by growing eight inbred strains of corn and their F_1 hybrids in a controlled environment with nutrient solutions containing different levels of nitrate. The question is whether increased efficiency in the processes of nitrate uptake and/or utilization could account for heterosis in any of these hybrids. Each set of inbred strains and their hybrids were fertilized with a nutrient solution containing either no nitrate or reduced levels of nitrate. The height, fresh weight and dry weight of the aerial portion of the plants were measured after about 60 days of growth; the ratios between the value for the hybrid and the value for the better of the two parental strains were used as measures of heterosis. All four hybrids surpassed the better parent for each of the traits measured. In three of four cases, a portion of the heterotic response disappeared (as predicted) when nitrate was removed from the medium. One of the four hybrids showed increased heterosis at low levels of nitrate, indicating, according to the hypothesis, that increased efficiency of uptake and/or assimilation of nitrate contributes in part to this particular heterosis. (Heterosis almost certainly involves alterations in more than one biological process; however, this approach can only detect them one at a time.) An attempt was made with this interesting hybrid to localize further the heterotic effect within the nitrate assimilation pathway by using the intermediates, nitrite, ammonium and L-glutamine, as nitrogen sources. Alterations in the nitrate transport system appear to play a role in the heterotic effect since there were distinct differences found between supplementation with nitrate and the other nitrogen sources.

Since whole plant phenotypes are cumbersome and difficult to manipulate in growth chambers, we have characterized an in vitro root culture system for the inbreds and hybrid. The relative growth rates of roots from the inbreds and the hybrids in culture on various levels of nitrate are similar to the growth rates of plants of the identical genotypes; i.e., heterosis is expressed in the root cultures. The relative extent of heterosis was accentuated at lower nitrate concentrations; the root culture system provides a response identical to that observed in the whole plant. Root cultures were utilized to investigate the mode of inheritance of the response to stress levels of nitrate. Our preliminary conclusion is that response to nitrate stress is controlled by (at least) a single genetic locus. In heterozygotes, the two alleles at this locus permit more efficient uptake/utilization of nitrate than in either homozygote.

Root cultures were taken to examine the biochemistry of the characteristic. Both nitrate reductase and nitrite reductase were assayed in roots cultured on 20 μM nitrate. There was not a significant difference among the three root genotypes in these enzymes. Nitrate reductase activity averaged 200 nmoles/hr/mg protein, while nitrite reductase averaged 400 μmoles/hr/mg protein. There was, however, a distinct difference between the parental inbreds and the hybrid genotypes in the rate of nitrate uptake. Roots of the hybrid appear better able to take up nitrate from dilute solutions than are roots of the inbred parental genotypes. The preliminary analysis of this phenomenon indicates that the V_{max} is similar in all three genotypes (both inbreds and the hybrid in concentrations of nitrate up to 200 μM (V_{max} = 9.0 to 9.5 μmoles/gm/hr).

However, the K_m values are different for the three genotypes (inbreds, K_m = 23-24 x 10^{-6}M; while for the hybrid K_m = 14 x 10^{-6}M). Assays of backcross progeny indicate that the heterotic growth response to nitrate stress is completely associated with the lower K_m of the heterozygote.

The results of this work are not surprising: 1) heterosis can co-occur, in some instances, with increased efficiency for nitrate uptake; and 2) the metabolic processes responsible for heterosis are different in different hybrid genotypes. These experiments do show this to be a productive approach for the identification of specific processes contributing to hybrid vigor and for determining the individual biochemical components which contribute to yield (Carlson, unpublished).

In this example with nitrate, a more productive plant may be one that can better obtain nitrate from its environment. It is important to note that these investigations with hybrids will not directly determine the mechanism of heterosis in any particular instance. They serve only to identify the metabolic site which is important for greater productivity in a given environment and to indicate the quantitative nature of the change at that site.

Once the individual biochemical components of heterosis have been defined, it will be possible to utilize the single gene modification schemes required by cellular manipulation and by DNA modification to effect potentially positive agronomic alternatives.

ENVOI

Discovering the eventual role of molecular biology and the new genetics in crop production requires an approach to the classical, holistic descriptions of plant productivity with reductionist and analytic tools, bringing a number of traditionally disparate biological disciplines to bear on the unique and complex problems of agricultural plant biology and of plant breeding. This process has begun, and several different areas in agricultural plant biology are acting as sites of initiation for the scientific saga. I see the importance of this beginning not in the content of quantitative information which results, but in the distinct qualitative shift in approach. As the work toward a molecular understanding of crop production proceeds, it is important to remember that the responsibility for the utility of this information lies with the basic scientist.

Applause will not be forthcoming from the real world until this intellectual gingerbred results in real world consequences. Those amber, ochre, and opal waves of grain are still just a twinkle in

the genetic engineer's eye... and that is no nonsense!! The new
genetics is not only an ornamental for it <u>can</u> and has already been
utilized in plant breeding programs. However, its application demands discrimination, an awareness of the needs of plant improvement programs, and an active collaboration between plant breeders
and molecular biologists. Those who would harness new genetic technologies for the third world must understand the framework within
which the innovation will function. We must continue the process
of communication between the scientific and non-scientific disciplines if this goal is to be met.

ACKNOWLEDGMENTS

I thank a number of friends and colleagues for their assistance
during the preparation of this paper. Drs. N. Christianson and
S. McCormick and Ms. C. Hoisington each provided important contributions to the final form. Naturally, I take full responsibility
for the results!

REFERENCES

Baker, L., Chen, N.C., Parrot, J., Tai, W., and Jacobs, T., (un-
    published data).
Bhojwani, S.S., Evans, P.K., and Cooking, E.C., 1977, Protoplast
    Technology in Relation to Crop Plants: Progress and Problems,
    Euphytica 26:343-360.
Bottino, P.J., 1975, The Potential of Genetic Manipulation in Plant
    Cell Cultures for Plant Breeding, Rad. Bot. 15:1-16.
Brown, A.W.A., Byerly, T.C., Gibbs, M., SanPietro, A. (eds.), 1975,
    Crop Productivity - Research Imperatives. Michigan Agr. Exper.
    Sta. and C.F. Kettering Foundation.
Carlson, P.S., (unpublished data).
Carlson, P.S., and Polacco, J.C., 1975, Plant Cell Cultures: Genetic
    Aspects of Crop Improvement, Science 188:622-625.
Han, Hu, Tze-ying, Hsi, Chun-Chow, Tseng, Tsun-wen, Ouyand, Chien-
    kang, Chang, 1978, Application of Anther Culture to Crop Plants,
    <u>in</u>: Frontier of Plant Tissue Culture 1978, T.A. Thorpe, ed.,
    University of Calgary Press.
Heinz, D.J., Krishnawurth, M., Nickell, L.G., and Maretzki, A., 1977,
    Cell, Tissue and Organ Culture in Sugarcane Improvement, <u>in</u>:
    Plant Cell, Tissue and Organ Culture, eds., J. Reinert and
    Y.P.S. Bajaj, Springer-Verlag, Berlin.
Kleinhofs, A., and Behki, R., 1977, Prospects for Plant Genome
    Modification by Non-conventional Methods, Ann. Rev. Genet.
    11:79-101.
Orton, T. (unpublished data).

Thomas, E., King, P.J., and Potrykus, I., 1979, Improvement of Crop
    Species via Single Cells in vitro - An Assessment, Z.
    Pflanzenzuecht.82:1-30.
World Food and Nutrition Study, 1977, National Academy of Sciences,
    Washington, D.C.

BIOMASS PRODUCTION AND UTILIZATION

H.T. Huang and L.G. Mayfield

National Science Foundation*

Washington, D.C. 20550

The theme of this Symposium is the linking of basic research
to crop improvement programs for the less developed countries.
The participants are asked to analyze the needs for advanced technology in developmental agricultural programs in the less developed
countries (LDC's) and examine the nature of the contributions to
be expected from basic research laboratories in the developed
countries in such areas as biomass production, nitrogen fixation,
plant breeding, etc. We are, furthermore, urged to explore ways
by which basic research may be more effectively linked to the
developmental process, to discuss the training of researchers that
might be needed to improve this linkage, and to determine how
funding resources should be divided between basic research (i.e.,
long-term projects) and applied research and development (i.e.
short-term projects).

NATURE OF LINKAGE AND CONSEQUENCES

We would like to comment briefly on these general issues before
moving on to our assigned area of Biomass Production and consider
a few specific research activities in this area in the United States
which may be relevant to the developmental process in the LDC's.
It seems to us that our main concern in this symposium is with the
linkage process between basic research, as practiced in the developed
countries, and crop improvement programs in the LDC's. What exactly

---

*Any opinions, findings, and conclusions or recommendations expressed
in this paper are those of the authors and do not necessarily
reflect the views of the National Science Foundation.

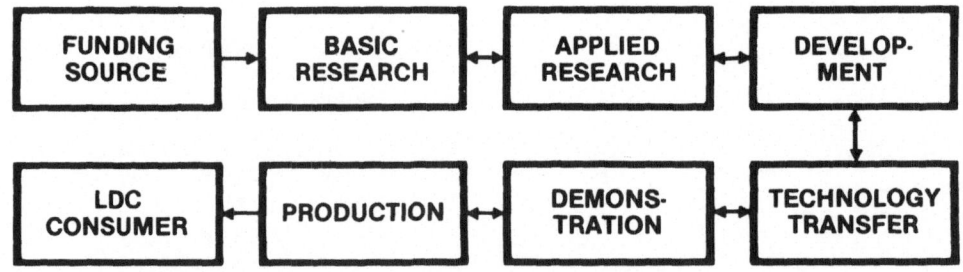

LINKAGE BETWEEN BASIC RESEARCH AND AGRICULTURAL PRODUCTION
AFTER J. A. MORTON IN "TECHNOLOGY TRANSFER AND INNOVATION", NSF 67-5

Figure 1.   Model of linkages which may exist between the less
            developed countries (LDC) and basic research for crop
            production.   (Source, NSF).

is the nature of the linkage? Our own perception of this linkage
is depicted by the model shown in Figure 1. It is a modification
of a model used to describe the linkage between basic research and
commercialization activities at the Bell Telephone Laboratories
(1), which has had a highly successful record in bringing the results
of basic research to practical fruition. In the context of this
symposium, we may regard the activities on the upper portion of
the diagram as those which occur primarily in the developed countries, and the activities on the lower portion as those which
occur primarily in the LDC's. We see at once that the linkage is
quite indirect and depends on connections forged through several
intermediate elements, such as applied research, development,
technology transfer, etc. All the intermediate links must be
successfully completed before a breakthrough in basic research
will produce a concrete improvement in crop productivity.

     Just as a chain is only as strong as its weakest link, so the
linkage between basic research and crop productivity improvement
is only as effective as the weakest element in the overall linkage
process. Thus, to ensure that an effective linkage is maintained,
it is imperative that a healthy basic research effort is supported
adequately by a correspondingly healthy effort in applied research
and development. In other words, one approach to improving the
linkage is to provide a balanced funding support for basic research,
applied research and development programs so that a scientific
discovery in basic research, if it has the potential, will be
expeditiously translated into usable technology. The technology
in turn has to be transferred, adopted and utilized before it can
have an impact on crop improvement. This is a slow process. It
has been estimated, in a recent study conducted by the National
Association of State Universities and Land Grant Colleges, that
it takes fourteen years before a basic research discovery will
materialize in the form of practical benefits in agricultural
productivity (2). The benefits usually peak at about twenty years

and then decline. For most LDC's, we would venture to guess that
this is too long a time to wait for a significant improvement in
crop productivity. Thus, it is not surprising that LDC's tend to
be much more interested in technology transfer than in basic research (3). For some time to come, we can expect LDC's to be largely
dependent on the developed countries for the fruits of basic research,
applied research, and development.

Technology, however, is not a static entity. In the course of
its transfer from a developed country to a LDC, it may have to
undergo substantial changes in order to overcome the social,
political and environmental barriers to its adoption in a new home.
Basic research can often play a significant role to provide the
information and insight needed to facilitate the transfer process.
Thus, another approach to improving the linkage between basic
research and crop productivity is to encourage basic researchers
to become interested and participate in technology transfer,
applied research and development activities. Their involvement
will not only strengthen these other links of the chain but will also
provide them with a better preception of the type of basic discoveries which will have a significant relevance to practical crop
improvement programs in the LDC's. This concept can be applied
towards the training of researchers in basic and applied research
oriented towards plant productivity problems. It is, of course,
an approach which has already been practiced for years by the
Boyce Thompson Institute, whose personnel have participated in all
phases of activity in this linkage, ranging from basic research to
technology transfer. It would be instructive to hear the views
of our hosts on the effectiveness of this concept and the problems
associated with its implementation.

FUNDING ALLOCATION FOR AGRICULTURAL R AND D

We have said earlier that to maintain an effective linkage it
is imperative to have a balanced distribution of funding support
for basic research, applied research and development activities.
But we have no objective guidelines to determine what constitutes
a proper balance. The organizer of this symposium, Dr. Staples,
has alluded to the difficulties involved in such a task at some
length in his letters of invitation to participants in this
symposium.   He has referred to past efforts to throw light on this
issue in the realm of weapons development conducted by the Department
of Defense (4) and in biomedical research by Comroe and Dripps (5).
The former concluded that basic research is less effective than
"targeted" research, while the latter concluded that basic research
pays off in terms of key discoveries at nearly twice the level of
applied research and development combined.

Both studies, however, skirt the key issue of what actually constitutes an appropriate distribution of limited funding resources among basic research, applied research and development. This is the information that the policy maker has to have in order to formulate an equitable and effective science policy in terms of national needs. Considering the intrinsic character of the activities, we would expect development to be more expensive than applied research, which, in turn, is most likely to be more expensive than basic research. Thus one would expect, in a general way, in a technologically developed country, total expenditures would decrease in the order of development, applied research and basic research. For the United States this is indeed the case. The information we need is readily available in a recent publication from the National Science Foundation (6). As shown in Figure 2, which summarizes the expenditures for R and D estimated for fiscal year 1979, basic research accounts for 13%, applied research 23%, and development 64% of the total national effort. The expenditure for applied research is slightly less than twice that of basic research and the expenditure for development is slightly less than three times that for applied research. These ratios have been held more or less constant during the past ten years as indicated in Figure 3. The Federal Government is the principal funding source for basic research, while the Federal Government and industry share almost equally in the support of development work.

These allocations are the sum total of multitudes of individual decisions and compromises in the Federal agencies, in Congress and

Source of Funds	Dollars in Billions		
	Basic Research	Applied Research	Develop- ment
Federal Gov't	4.19	5.29	14.34
Industry	0.90	4.85	16.04
Universities	0.60	0.34	0.06
Others	0.36	0.25	0.09
Total	6.05	10.73	30.53

Funding for Research & Development, U.S.A. 1978 (estimated)
Source: NSF 78-313

Figure 2.   Summary of funding for research and development estimated for fiscal year 1979. (Source: NSF).

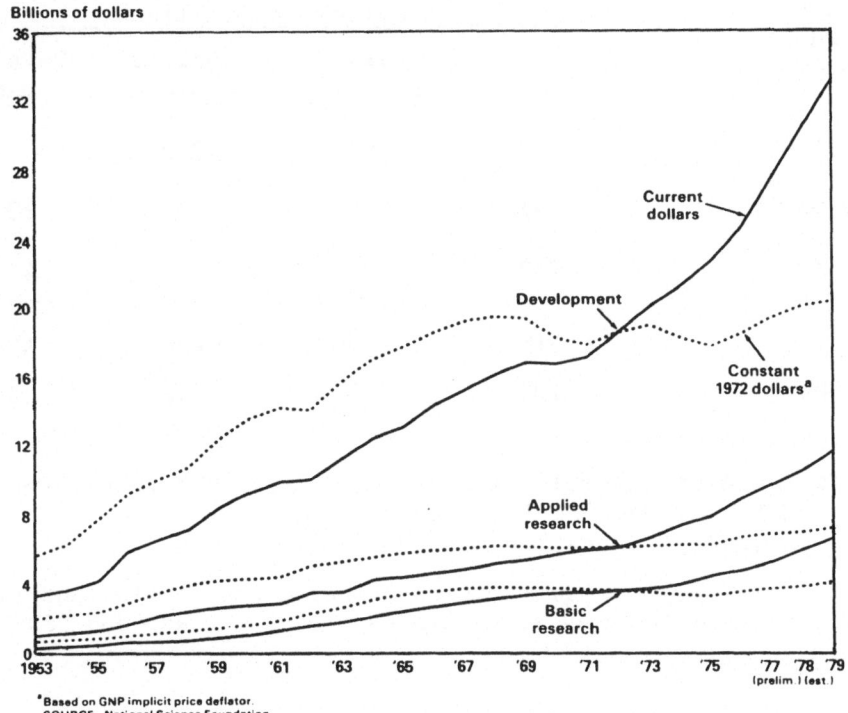

Billions of dollars

Figure 3.   National spending for research and development by
            character of work: 1953–1979. (a) based on gross
            national product implicit price deflator. (Source, NSF).

in the major industrial companies. They do not necessarily reflect
the conscious execution of a sane and effective national policy.
Nor do we have any assurance that this particular distribution is
the best and most effective way of allocating the limited resources
available to the total R and D effort. Current concern for the lack
of productivity growth in the U.S. at least signals the possibility
that the funding pattern for research and development is not for-
mulated in the most effective way (7).

     Having ascertained the overall national pattern of funding
allocation, we can now take a look at how closely this pattern
holds for R and D in the field of agricultural technology. From
two recent surveys conducted by the National Science Foundation
(8, 9), and information provided by the Science and Education
Administration of the U.S.D.A. (10), we have arrived at the estimated expenditures for fiscal year 1977 shown in Figure 4. The
industry estimates are based on data available for three industries: agricultural chemicals, farm machinery and food and kindred
products. These numbers indicate that in agricultural R and D,

Source of Funds	Dollars in Millions			
	Total	Basic Research	Applied Research	Development
Federal Gov't	547	204	320	23
State Gov't	341	136	102	103
Industry	723	39	252	432
Total	1,611	379	674	558
% of total	100	23	32	35

Funding for Agricultural R & D in the U.S., 1977 (estimated)
Source: National Science Foundation
       Science & Education Administration, USDA

Figure 4.   Funding for agricultural research and development in
            the U.S. in 1977, estimated. (Source, NSF and USDA).

basic research is supported at almost twice the level and development at only about half the level estimated for the nation as a
whole. This could be interpreted to mean that in agriculture
development is underfunded. They could also mean that many of the
results of basic research are not ready yet for development and
application. We hope these issues will be examined in depth in
the policy session of this symposium.

But we would like to end this part of our paper by pointing
out one more conclusion that can be drawn from Figure 4, and that
is the predominant role of industry in the execution of development
work. Unlike the total R and D effort of the nation, in the
agricultural sector the Federal and State governments contribute
only about one-fifth of the expenditures for development. Thus,
it is possible that certain innovative technologies which do not
readily lend themselves to the development of marketable products,
for example, the biological control of pests, are not receiving
adequate support to bring them expeditiously to practical application and utilization.

BIOMASS CONVERSION

We will now return to the specific topic of this paper,
namely, Biomass Production, or more appropriately, Biomass

Production and Utilization, since in this paper we will be concerned with utilization as much as with production of biomass. You have heard from our eminent colleague, Dr. Zelitch, a number of exciting ideas for improving crop productivity based on our understanding of the fundamental process for biomass synthesis in plants. They represent long-term projects where utilization is some years away. We will now report on several types of investigations on biomass utilization and production which are of the short or intermediate term variety where a payoff may be expected in five to ten years. Success in these applied research ventures will no doubt accelerate the need for and justify the support of further expanded long-term basic research activities. In this sense, these ventures will, by virtue of their very existence, strengthen the linkage between basic research and crop productivity.

The overall theme of these applied research activities is the utilization of unutilized or underutilized renewable resources. The immediate driving force behind them is the quadrupling of petroleum prices, coupled with the oil embargo of 1973. This event, shown graphically in Fig. 5, signaled for the U.S. the end of the age of cheap energy. It led to drastic increases in the cost and spot scarcities of fuel and materials such as chemical feedstocks, synthetic polymers and nitrogen fertilizers, all essential to the smooth functioning of a modern technological society.

As the supply of petroleum stabilized in the aftermath of the oil embargo of 1973 thru 1974, it became clear that the increasing dependence of the U.S. economy on imported petroleum is not a desirable strategy. Petroleum is a nonrenewable fossil resource, and an increasing share of U.S. needs is imported. One obvious way to alleviate the nation's dependence on foreign supplies is to replace them, as much as possible, by alternative resources which are renewable and of domestic origin.

Biomass, in the form of lignocellulosic residues, appears to be an attractive alternative to petroleum and other fossil energy resources such as natural gas, oil shale and coal. A good deal of lignocellulosic materials are already available as residues from agricultural and forestry operations and municiple wastes. According to a study conducted by the Committee on Renewable Resources for Industrial Materials (CORRIM) of the National Academy of Sciences (11), the amount of such materials generated in the U.S. each year is in the neighborhood of 2 billion tons. If we deduct the hardwoods, we would still have 800 million dry tons that are either already collected, e.g., urban waste, manure and manufacturing residues, or in an easily collectable state, e.g., crop residues.

Since 1976 we have had at the National Science Foundation, initially under the RANN (Research Applied to National Needs)

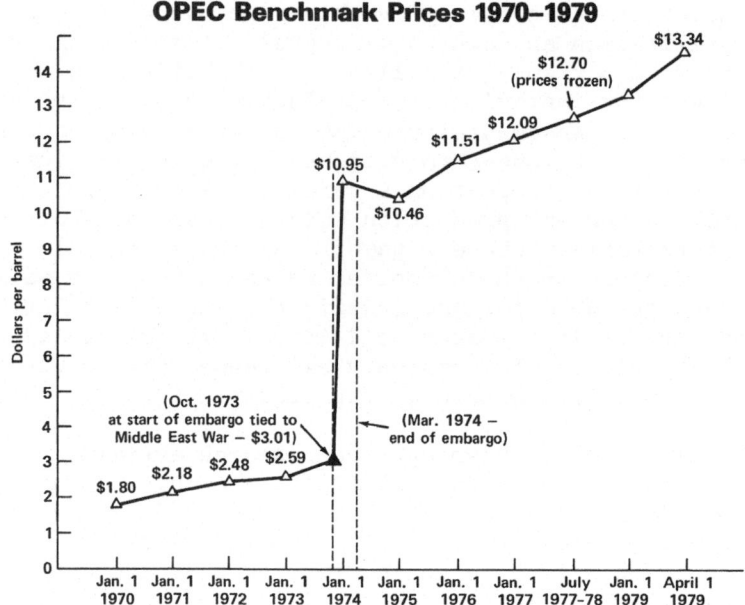

Figure 5. OPEC Benchmark prices, 1970-1979.

Directorate, and currently under the ASRA (Applied Science and
Research Applications) Directorate, a program aimed at the possible
replacement of petroleum, and natural gas, by biomass residues for
the production of chemical feedstocks and other critical industrial
materials. The program is called Alternative Biological Sources
of Materials (ABSM) and currently consists of two major thrusts.
The first is to develop efficient and energy conserving technologies
to convert lignocellulosic biomass into chemical feedstocks. The
second is to produce critical materials from plants which can be
grown on marginal lands in the more arid regions of the United
States.

     We will take up Biomass Conversion first. The CORRIM study
(12) has shown that of the 18 million tons of synthetic polymers
produced in the U.S. each year, 95% are derivable from ethylene,
butadiene and phenol. Ethylene and butadiene are obtainable from
ethanol, which can be produced from glucose, the product of the
hydrolysis of cellulose. Phenol and related compounds can be
obtained from hydrogenation or hydrogenolysis of lignin. The CORRIM
study has shown that 58.5 million tons of lignocellulose will be
needed to produce the 17.5 million tons of polymers currently being
derived from petroleum feedstocks.

     Now, the hydrolysis of cellulose in wood by acid catalysis to
simple sugars has actually been known for a long time. It was

practiced on a commercial scale in the U.S. during World War I, in Germany during World War II, and is still being practiced in the USSR today. A complete conversion of lignocellulose to useful feedstocks can be visualized in the flow diagram shown on Fig. 6. Essentially, there are three operations involved in the process:

1. Hydrolysis of hemicullulose to pentose;
2. Hydrolysis of cellulose to glucose;
3. Conversion of lignin to aromatic chemicals.

Operation 1, the hydrolysis of hemicellulose to the monomeric pentoses, can be achieved in good, i.e., about 90% yield. Operation 2, the hydrolysis of cellulose, is made difficult by the crystallinity of the substrate and its intimate structural organization with lignin, which tends to restrict the accessibility of acid to the bonds to be cleaved. Thus, conventional technology requires the use of high temperature and acid concentrations that cause decomposition of the resultant monomeric hexoses. The yield of fermentable glucose is only 50%. Similarly, operation 3 is also

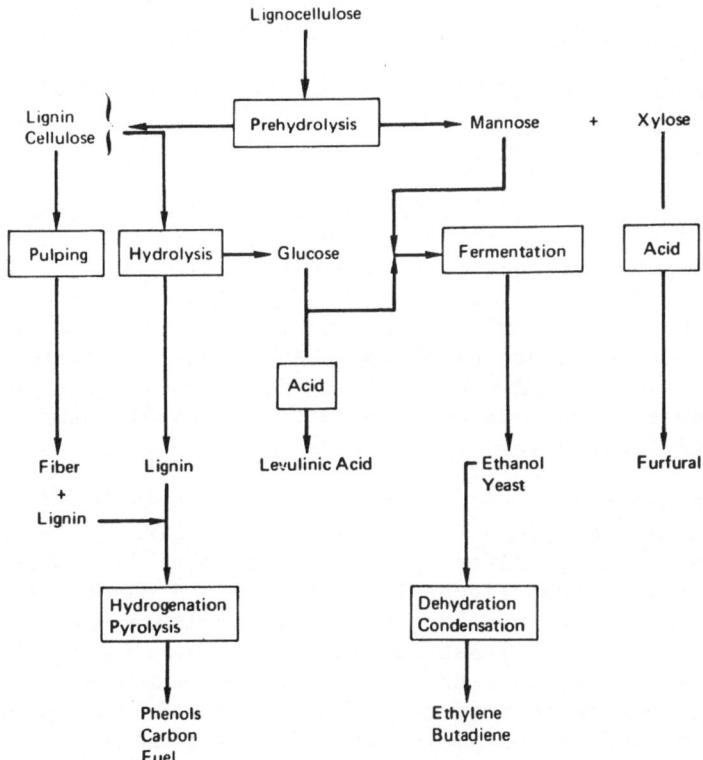

Figure 6.   Schematic flow chart of a sample lignocellulosic chemical plant.

restricted by the difficulty of separating the lignin of high
quality from cellulose, although to a limited extent this is already
being done in the manufacture of paper pulp from wood.

The NSF program has focused on innovative approaches to improve
the efficiency of all three operations, especially those based on
biological or enzymatic energy conserving processes. In the
hydrolysis of hemicellulose, an effort is being made to produce,
isolate and purify enzymes which hydrolyse xylan, the predominant
hemicellulose in grasses, cereal grains and hardwoods, into xylose
and small amounts of other sugars (13). Xylose is fermentable to
ethanol, or it can be hydrogenated to xylitol, a potent sweetener,
already in use in Europe.

For the breakdown of cellulose, two types of investigations are
being supported. The first deals with methods of pretreatment of
the lignocellulose, so as to render the cellulose more amenable to
both enzymatic or acid hydrolysis (14, 15, 16). Certain metal
complexes, such as cadoxen, a solution of cadmium oxide in 28
percent aqueous ehtylenediamine, readily dissolve cellulose in
lignocellulose, and the reprecipitated cellulose is hydrolyzed by
the cellulase enzymes of Trichoderma reesei to give >90% yield of
glucose (17). The second approach is the controlled pyrolysis of
lignocellulose to levoglucosan, a glucose anhydride which is easily
hydrolyzed to glucose when it is exposed to water (18). Conditions
are being sought which would give high yields of levoglucosan from
lignocellulose.

In the conversion of lignin the emphasis is to transform lignin
by the use of microbial enzymes to industrially useful materials,
in as high a polymeric state as possible (19, 20, 21). The objectives include the isolation and identification of lignin transforming organisms (bacteria, yeast and fungi) that live in symbiotic
association with wood attacking insects, as well as free living
lignolytic organisms. A number of lignin degrading actinomycetes
have been isolated and are being mutated to select mutants of
superior activity (19).

The NSF program is coordinated with a Department of Energy
program aimed at the production of Fuels from Biomass (FFB) in
which a principal objective is the production of liquid fuel, that
is to say, ethanol from lignocellulose (22). A major component of
the FFB program is the development of silviculture energy farms.
It is estimated that at least 30 million of the nation's 1.3 billion
acres of grassland and forest can be readily made available for
this approach. Biomass productivity under close-spaced, short
rotation conditions is estimated at 5-13 dry tons per acre per year
with current technology, depending upon species and site selection.
A 1,000 acre silviculture plantation is being established to test

this concept, and an award has been made to build and operate a pilot plant for the conversion of soft wood biomass into ethanol (23). The process will be integrated with the production of naval stores, an extractive from pines which can be readily removed by steam distillation before the wood is fed into the conversion process. A principal reason for the choice of this system is the observation that the content of oleoresins in pines can be dramatically increased by the injection of a small amount of paraquat, a herbicide, into the cambium of the tree (24). Oleoresin content as much as 20 percent of the total dry weight of the tree has been achieved by this procedure. Integration of the production of extractives with the conversion of biomass into fuels and industrial materials certainly offers an attractive way to bring the whole concept into the realm of economic feasibility.

The production of biomass by silviculture and conversion of biomass to fuels and chemicals are technologies that can have a profound impact on the economic development and well being of many LDC's. According to a recent WorldWatch paper, the shortage of firewood has reached crisis proportions for more than a third of the world's people, in the densely populated Indian subcontinent, in the semi-arid regions fringing the Sahara Desert and in parts of Latin America (25). Improved silviculture management systems can reap benefits within a short period of time if the affected LDC's are willing and able to put them into practice. For other countries, which are blessed with abundant renewable resources, such as Brazil, the conversion of biomass into fuel can greatly reduce their dependence on imported petroleum as sources of fuel and chemical feedstock. Indeed, the production of alcohol from sugar cane and cassava, for use as automotive fuel, is already an established component of the national energy policy in Brazil (26). In the United States, the recent publicity on gasohol (27), a blend of 10% ethanol and 90% gasoline, cannot help but stimulate further interest in ethanol from biomass as a factor in the energy future of the nation.

ARID LAND PLANTS

From the extraction of naval stores from pines it is conceptually but a small step to the extracton of other useful materials and chemicals from other plants. In the second major element of NSF's Alternative Biological Sources of Materials (ABSM) program, we have taken such a step and have further narrowed our focus on the production of critical materials from as yet undomesticated plants which are particularly suitable for cultivation in the arid and semi-arid regions of southwestern United States. We are currently supporting research on the domestication and utilization of three plants native to the arid regions of North America, which

have aroused unusual interest as potential crops for the underutilized marginal lands of southwestern U.S. and northern Mexico.
They are guayule, jojoba and the buffalo gourd. All three are well
known to and have been used in a limited way by the native Indian
communities of the region.

1.   Guayule

     Guayule, Parthenium argentatum Gray, is a low shrub native to
the arid regions of north central Mexico and the big bend area of
western Texas. The specific locations where native stands are
found are shown in Figure 7. The stems and roots of the plant
contain a rubber which, when purified, is virtually indistinguishable
from the natural rubber of the familiar rubber tree, Hevea
brasiliensis.

     Natural rubber from Hevea brasiliensis is today a major and
vital commodity in the U.S. economy. It is indispensable for bus,
truck and airplane tires, and for steel belted radial tires for
automobiles. Each year the U.S. imports 750,000 tons at a cost of
$650 million. Eighty percent of the world's production resides in
three countries, Maylaysia, Indonesia and Thailand in southeast
Asia. It is believed that in the 1980's the world's demand for
natural rubber will exceed the supply from Hevea Plantations,
thereby resulting in a shortfall.

     During World War II, when supplies of Hevea rubber from
southeast Asia were cut off, the U.S. Government established an
Emergency Rubber Project (ERP) which in 3 1/2 years planted 32,000
acres of guayule at 13 sites in three states. It produced one
billion seedlings and 3 million pounds of resinous rubber for the
war effort. Towards the end of the project, 15 tons of rubber
were produced daily in factories in Salinas and Bakersfield,
California.

     This massive effort was abandoned after World War II when
synthetic rubber became a commercial reality and Hevea rubber from
southeast Asia again became readily available. In 1946, 30,000
acres of guayule in plantations were burned or disced in the ground,
with the destruction of about 10,000 tons of natural rubber.

     And yet, thirty years later, an ad hoc panel of the National
Academy of Sciences, after an extensive reevaluation of the
existing situation, again recommended that a national effort be
launched to develop guayule as a U.S. domestic source of natural
rubber (28). In the meantime, Mexico had already decided to start
a new guayule program and had built a pilot plant to process natural
guayule stands in Saltillo, Coahuila, with a capacity of handling
one ton of shrubs per day. The deresinated rubber has been tested

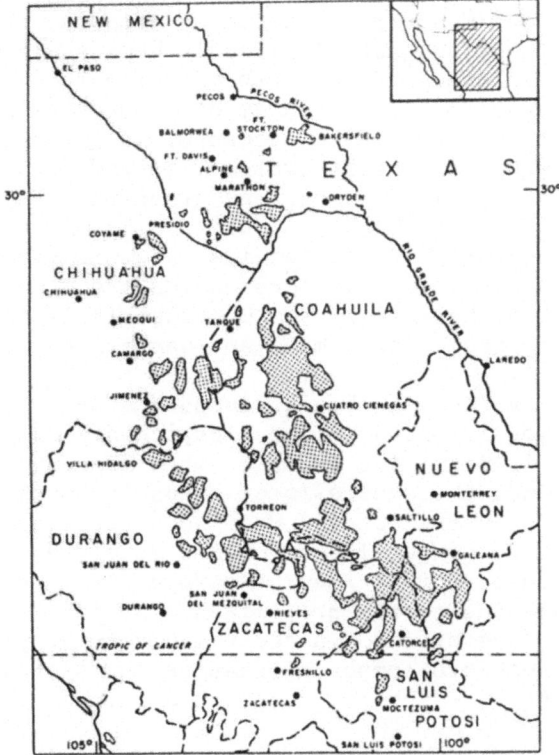

Figure 7. Distribution of native guayule in Mexico and Texas.

extensively (28) and found to be equal in quality to current commercial _Hevea_ rubber.

    NSF Support for guayule research started in 1976 and has
followed closely the recommendations of the NAS panel. Funding
has been provided for:

    o   Seed Collection (29)
    o   Plant Breeding and Experimental Plantings (30)
    o   Rubber Quality Evaluation (31)
    o   Tissue Culture (32)
    o   Dermatotoxicity (33)
    o   Feasibility Study (34)
    o   Technology Assessment and Environmental Impact (35).

    We cannot discuss these projects separately, but we can tell
you about a novel approach to improve rubber yield currently supported by the National Science Foundation which was not included in
the NAS recommendation. It is a project conducted jointly by the
USDA Fruit and Vegetable Chemistry Laboratory in Pasadena and the

Los Angeles State and Country Arboretum in Arcadia, California
(36). It came into being quite indirectly because of Henry Yokoyama's
interest in the regulation of biosynthesis of carotenoid pigments
in fruits and vegetables. The pathway of biosynthesis of the red
carotenoid pigment of the tomato, lycopene, is presented in Figure
8. One of the precursors, zeta-carotene, is the principal yellow
pigment found in the peel of lemons and grapefruits.

    You will note that the carotenoids shown in Figure 8 are all
of the _trans_ configuration. _Cis_ carotenoids are also known to
exist in nature, but they are rare. One example is _cis_ lycopene,
which occurs in the tangerine tomato and has a light orange color.

    Yokoyama discovered a number of compounds which derepress the
synthesis of enzymes involved in the biosynthesis of _trans_ carotenoids in citrus fruits. For example, when a lemon is treated
with such a compound, the synthesis of _trans_ lycopene is enhanced
and it develops a deep red color. He also discovered a compound
which derepresses the synthesis of enzymes involved in the biosynthesis of _cis_ carotenoids. When he sprayed it on a lemon, the
lemon turned orange due to the predominance of _cis_ lycopene. These
compounds are a series of chlorinated phenoxy-triethylamines
(Figure 9). The 4-chloro compound enhances the synthesis of _trans_
lycopene, while the 3,4-dichloro compound activates the synthesis
of _cis_ lycopene.

    Carotenoids and natural rubber both belong to the same family
of products derived from isoprene. Natural rubber is a _cis_ polymer
of isoprene. When Yokoyama became aware of the revived interest

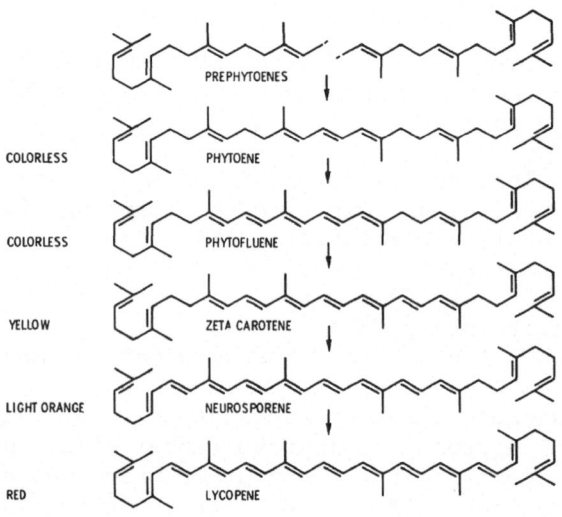

Figure 8. Biosynthesis of carotenoids.

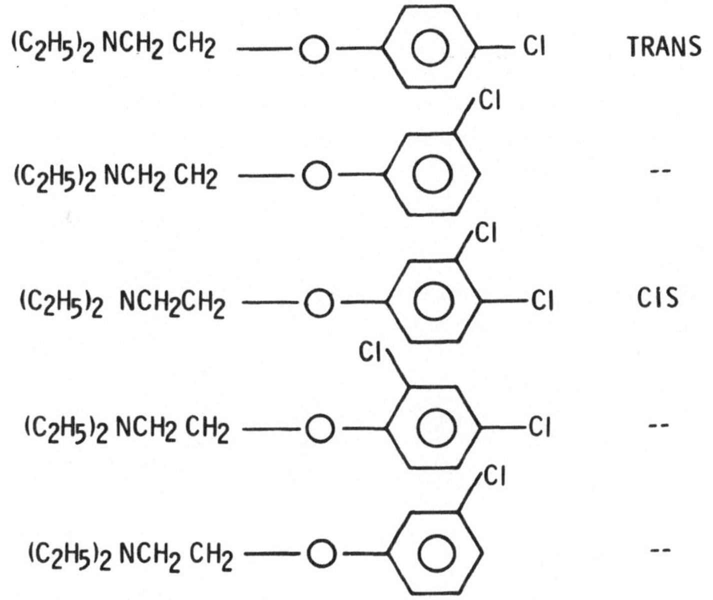

$(C_2H_5)_2\,NCH_2\,CH_2$ ——O—⟨O⟩—Cl   TRANS

$(C_2H_5)_2\,NCH_2\,CH_2$ ——O—⟨O⟩   --

$(C_2H_5)_2\,NCH_2CH_2$ ——O—⟨O⟩—Cl   CIS

$(C_2H_5)_2\,NCH_2\,CH_2$ ——O—⟨O⟩—Cl   --

$(C_2H_5)_2\,NCH_2\,CH_2$ ——O—⟨O⟩   --

Figure 9.   Bioregulatory activity of chlorinated phenoxy-
            triethylamines.

in guayule, he managed to persuade George Hanson at the Aboretum
to give him a number of young guayule plants for testing with these
compounds.   The trans activator was inactive.   Those inactive on
citrus fruits were also inactive on guayule.   However, the cis
activator increased the rubber content from 2- to 3-fold.

     Yokoyama and Hanson are now continuing these studies on field-
grown plants.   In addition to determining the optimum parameters
for applying the activator, they will also 1) attempt to develop
an activator which does not contain chlorine and 2) integrate
response to chemical activators into the plant breeding program.

     In 1942, the ERP surveyed southwestern United States and
identified 5 million acres of land in California, Arizona, New
Mexico and Texas as suitable for guayule cultivation (Figure 10).
Whether it can be grown in the arid regions successfully as a crop
without irrigation is by no means clear, but there is no question
that the amount of water it needs annually is less than that for
most other traditional crops (Figure 11).   Thus, it is not surprising that the representatives of the four southwest border
states have been able to mobilize sufficient political momentum
during the past year to effect passage of a bill in Congress,
"Native Latex Commercialization and Economic Development Act of
1978" (PL 95-592) which authorizes the establishment of a Joint
Commission, with the Department of Agriculture and the Department
of Commerce as the lead agencies, to develop guayule as a domestic
source of natural rubber.   Funds up to $30 million in four years
have been authorized for this purpose.   As the Joint Commission

Figure 10.   Areas in the United States with climate considered
                     suitable for the cultivation of guayule as reported
                     by the Emergency Rubber Project, 1944 (28).

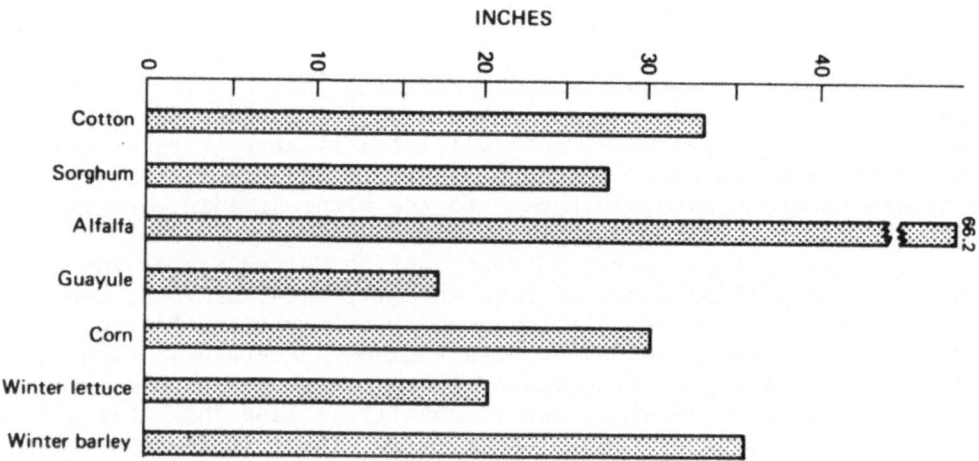

Figure 11.   Estimated consumption of water by guayule compared
                     with selected irrigated crops grown at El Paso, Texas.

becomes organized and fully operational, it will no doubt assume
responsibility for many of the projects now being funded by NSF
which are appropriately an integral part of the national program.

2.   JOJOBA

Jojoba, Simmondsia chinensis, (Link) Schneider, is a hardy
shrub that grows wild over large areas in the Sonoran Desert of
Arizona, California and Mexico. Although usually bushy, it may
reach 10 feet in height, offering a thick cover in the desert.
Its natural life span is estimated at between 100 and 200 years.
It tolerates extreme desert temperatures, with daily highs of 35
to 45 degrees centigrade. It is drought resistant and thrives
under soil and moisture conditions not suitable for most agricultural crops. However, for high productivity, 15 to 18 inches of
rainfall would be advantageous.

Jojoba bushes are either staminate (male) or pistillate
(female). It is not possible to distinguish the sex of a seedling
until it actually flowers. Only the female bears seed, but male
plants are needed to pollinate the female plants. Newly planted
seedlings require about 5 years before they start bearing seeds.
At present, existing plants are highly variable in amount of seed
produced. The average is 5 pounds per bush, with a high of 12
pounds and a low of 1.5 pounds.

The key to the current interest in jojoba lies in the oil,
which accounts for about 50% of the dry weight of the seed. Unlike
the triglycerides found in most oil seeds, jojoba oil is actually
a liquid wax, with a composition very similar to that of sperm
whale oil. It is composed entirely of esters of high molecular
weight. More than 85 percent of the esters are combinations of
C-20 and C-22 acids and alcohols. Most of the acids and alcohols
have their double bonds at C11 and C13, unlike other natural fats
or oil in which the double bond is usually at C9.

Sperm whale oil has been in great demand as a lubricant for
high pressure machinery. In 1970 the sperm whale was declared an
endangered species and its oil could not be imported into the
United States. Jojoba oil offers an attractive alternative to
sperm whale oil. In 1972 the Indian Division of the Office of
Economic Opportunity (OEO) initiated a program, later continued by
the Office of Native American Programs (ONAP) and by the Bureau of
Indian Affairs (BIA), aimed at the development of jojoba as an
agro-industry to improve the economic condition of the impoverished
Indian reservations in southwestern United States. The National
Academy of Sciences has issued two reports (37, 38) favoring the
development of jojoba as a cultivated crop for arid lands.

The BIA program has supported the collection of seeds from

wild plants on Indian reservations, the pressing of seeds for oil,
the distribution of oil for evaluation in potential applications
and the establishment of new plantations in Indian lands. NSF
support for jojoba research was started in 1977 and includes two
types of investigations.

The first deals with germplasm collection, manipulation of
native stands, plant breeding and selection of superior varieties
suitable for use in commercial plantations, at the University of
Arizona and at the University of California, Riverside (39, 40).
The Riverside project (41) also includes research on cytogenetics
and the regeneration of whole plants from cell culture.

The second type of investigation is aimed at the utilization
of the seed meal. After the removal of oil, the seed meal contains
about 30 percent protein and other materials. The amino acid
composition of the protein is given in Table I. The seed meal
would be a useful source of protein in animal feed if it did not
contain a number of toxic substances of which the principal components are simmondsin and simmondsin-2'-ferulate. Research is in
progress to develop effective and inexpensive methods to detoxify
the meal so that it can be used as a source of protein in the diet
of poultry, sheep and cattle (41).

Table I. Amino Acid Composition of Deoiled Jojoba Seed Meal

amino acids	Apache 377, %	SCJP 977, %
lysine	1.05	1.11
histidine	0.486	0.493
arginine	1.56	1.81
aspartic acid	2.18	3.11
threonine	1.14	1.22
serine	1.04	1.11
glutamic acid	2.40	2.79
proline	0.958	1.10
glycine	1.50	1.41
alanine	0.832	0.953
valine	1.10	1.19
methionine	0.186	0.210
isoleucine	0.777	0.866
leucine	1.46	1.57
tyrosine	1.04	1.05
phenylalanine	0.919	1.07
cystine and cysteine	0.791	0.519
tryptophan	0.492	0.559

3. BUFFALO GOURD

The last, but not the least, of this group of arid land plants
is the buffalo gourd, Cucurbita foetidissima. It is a perennial
xerophyte that grows wild in regions both east and west of the
Rockies. The useful life span is believed to be about 15 years.
Each season a plant produces extensive vine growth and as much as
200 fruits, or gourds. A gourd contains from 200 to 300 seeds.
The seed, which contains 30-40 percent edible oil and 30-35 percent
protein, is the primary crop. The oil compares favorably with other
common edible oils in terms of unsaturated fatty acid content.
The protein, like many plant proteins, is slightly low in methionine.

The vine itself may also represent a useful resource as feed.
When harvested prior to frost it has a protein content of 10 to
13 percent, and a rumen digestibility of 60 percent. The plant
also produces a large storage root containing about 15 percent starch
on a wet weight basis.

NSF is currently supporting a project at the University of
Arizona on the breeding, domestication and utilization of the
buffalo gourd (42). Some day, this "roadside weed", which many
ranchers have tried unsuccessfully to eradicate, may become a profitable commercial food crop for the southwestern United States.

We have now considered three potential crops, all native plants
already adapted to existing environmental and climatic conditions,
for marginal lands in the arid zones of the southwestern United
States. If they can be domesticated, cultivated, processed and
utilized, they will contribute in a major way to the economic welfare
of the region. But success in these endeavors will depend greatly
on how effectively and expeditiously basic and applied research,
particularly in the application of cell genetics to plant breeding,
and the stimulation of plant productivity by bioregulators, will
provide the technologies needed to increase the crop productivities
to a level which will make the plants economically viable.

Since vast areas of many LDC's lie in arid zones, establishment of these crops in the United States will undoubtedly stimulate
efforts to introduce them to the LDC's. The transfer of the
technologies involved will not be easy unless a favorable political
and social environment to reveive them is already in existence.
The experience we have had with the development of jojoba as a
crop for Indian reservations in the United States may be instructive.

From July 1971 to August 1978 the Federal Government has spent
almost $3 million in support of the Native American Jojoba Develop-
ment project via BIA (ONAP and OEO) to develop and commercialize

jojoba as a crop for Indian lands. Four Indian reservations
participated in the project, the Apache reservation in Arizona,
and the Cabazon, Pauma and Morongo reservations in California.
The effort was largely responsible for drawing attention of the
public to the unique qualities and potential market for jojoba oil.
By 1976 the potential of jojoba had become so well known that private entrepreneurs started to collect jojoba seeds, produce jojoba
oil and promote the establishment of jojoba plantations. Indeed,
by the end of 1978, 3,000 acres of jojoba have been planted by
private, non-Indian growers and only 220 acres planted by Indian
groups (43). These facts seem to speak for themselves on the
technology transfer mechanisms.

REFERENCES

1.   Morton, J.A., Proceedings of a Conference, "Technology Transfer
          and Innovation", National Science Foundation 67-5, p.21,
          1967.
2.   Guyford Stever, H. Office of Science and Technology Policy
          Memo to James T. Lynn, OMB, 1976.
3.   Benjamin, Milton R. "The New Global Mania: Transfer of
          Technology", "The Washington Post", Washington, D.C.,
          December 7, 1978.
4.   Sherwin, C.W. and R.S. Isenson, "Interim Report on Project
          Hindsight (Summary)", AD 642400, National Technical
          Information Service, Springfield, Va., October 31, 1966.
5.   Comroe, J.H., Jr. and R.D. Dripps, "Scientific Basis for the
          Support of Biomedical Science", Science 192, 105, 1976.
6.   "National Patterns of R and D Resources", National Science
          Foundation 78-313, 1978.
7.   Lepkowski, W., Chem. and Eng. News, p.14, Feb. 12, 1979.
8.   "Federal Funds for Research and Development", Vol. XXVII,
          Appendix C, National Science Foundation 78-312, 1978.
9.   "Research and Development in Industry, 1976", National Science
          Foundation 78-314, 1978.
10.   Fishback, Jane, John Fulkerson and Len Jensen, Private Com-
          munications.
11.   "Renewable Resources for Industrial Materials: A Report of
          the Committee on Renewable Resources for Industrial
          Materials", p.203, National Academy of Sciences, Washing-
          ton, D.C., 1976.
12.   "Renewable Resources for Industrial Materials: A Report of
          the Committee on Renewable Resources for Industrial
          Materials", p.202, National Academy of Sciences, Washing-
          ton, D.C., 1976.
13.   Reilly, P.J., "The Conversion of Agricultural by product to
          Sugars", National Science Foundation, PFR 77-00198, 1977.

14. Allen, B.R., "Pretreatment Methods for the Degradation of Lignin", National Science Foundation, PFR 77-25833, 1977.

15. Sarkanen, K.V., "Chemicals from Western Hardwoods and Agricultural Residues", National Science Foundation, PFR 77-08979.

16. Tsao, G.T., "Cellulose Hydrolysis with and without Pretreatment: Kinetics and Process Design", National Science Foundation, AER 76-11686.

17. Ladisch, M.R., C.M. Ladisch, and G.T. Tsao, Science 201, 743, 1978.

18. Shafizadeh, F., "Pyrolytic Conversion of Lignocellulosic Materials", National Science Foundation, AER 75-15930, 1975.

19. Crawford, D.L., "Conversion of Lignocellulose by Actinomycete Microorganisms", National Science Foundation, AER 75-23401, 1975.

20. Hall, P.L., "Enzymatic Transformation of Lignin", National Science Foundation, AER 76-11050, 1976.

21. Norris, D.M., "Isolation of Lignocellulose Transforming Microorganisms", National Science Foundation, PFR 77-08279, 1977.

22. "Program Summary, Fuels from Biomass Program", Department of Energy, DOE-ET-0022/1, January 1978.

23. O'Neill, D.J., "Design, Fabrication and Operation of a Biomass Fermentation Facility", Georgia Institute of Technology, Atlanta, Georgia, ET-78-C-01-3060.

24. "Renewable Resources for Industrial Materials: A Report of the Committee on Renewable Resources for Industrial Materials", p.234, National Academy of Sciences, Washington, D.C., 1976.

25. Eckholm, Erik, "The Other Energy Crisis: Firewood", Worldwatch Paper #1, Washington, D.C., September 1975.

26. Hammond, A.L., Science 200, 753, 1978.

27. Anderson, E.V., Chem. and Eng. News, p.8, July 31, 1978.

28. "Guayule: An Alternative Resource of Natural Rubber", National Academy of Sciences, Washington, D.C., 1977.

29. Rubis, D., "Natural Rubber from Guayule: Seed Collection and Study", National Science Foundation, AER 76-24666, 1976.

30. Hanson, G., "Breeding Improvement of Rubber Yield in Guayule", National Science Foundation, PFR 76-24472, 1976.

31. McIntyre, D., "Structure, Compounding, Physical and Rheological Properties of Guayule Rubber", National Science Foundation, INT 78-07481, 1978.

32. Murashige, T., "Development of Tissue Culture System for Guayule", National Science Foundation, PFR 78-25829, 1978.

33. Rodriquez, E., "Toxicological Studies on Guayule", National Science Foundation, PFR 78-25162, 1978.

34. Bragg, D., "Feasibility of Producing Natural Rubber from
       Guayule", National Science Foundation, PFR 78-12713,
       1978.
35. Foster, K., "A Technology Assessment of the Commercialization
       of Guayule", National Science Foundation, PRA 78-11632,
       1978.
36. Yokoyama, H., and G. Hanson, "Chemical Stimulation of Rubber
       Synthesis in Guayule", National Science Foundation, PFR
       78-09567, 1978.
37. "Products from Jojoba: A Promising New Crop for Arid Lands",
       National Academy of Sciences, Washington, D.C., 1975.
38. "Jojoba: Feasibility for Cultivation on Indian Reservations
       of the Sonoran Desert Region", National Academy of Sciences,
       Washington, D.C., 1977.
39. Hogan, LeMoyne, "Evaluation and Utilization of Native Jojoba",
       National Science Foundation, INT 76-18311, 1976.
40. Yermanos, D.M., "Development of Superior Cultivars of Jojoba",
       National Science Foundation, PFR 78-12709, 1978.
41. Verbiscar, A.J., "Jojoba Seed Meal as an Animal Feed", National
       Science Foundation, AER 76-23895, 1976.
42. Bemis, W.P., "Breeding, Domestication and Utilization of the
       Buffalo Gourd", National Science Foundation, AER 76-82387,
       1976.
43. Miller, William P. and Ned Anderson, "The Native American
       Jojoba Project", Paper presented at the Third International
       Conference on Jojoba, Riverside, Calif., Sept. 13-16, 1978.

BASIC RESEARCH IN BIOMASS PRODUCTION: SCIENTIFIC OPPORTUNITIES AND ORGANIZATIONAL CHALLENGES

Israel Zelitch

Department of Biochemistry
Connecticut Agricultural Experiment Station
New Haven, Connecticut   06504

Discussions on biomass production (the yield of plant dry matter) appear to have two separate elements, although both have much in common. These elements are primary food production, and the formation of biomass as a potential source of energy and basic chemicals. Most biomass is derived from the carbon dioxide fixed from the air during photosynthesis; and only about 5 to 10 per cent comes from minerals and nitrogen taken up from the soil. The biomass is largely determined by the gross photosynthetic CO_2 assimilation minus the losses by respiration. Crop yields, about which I am mostly concerned, are therefore the edible portion of biomass production.

The subject of obtaining fuels from biomass conversion will not be discussed, and only the primary production of biomass will be considered. Trees are usually thought of as the main source of biomass when fuel sources are suggested, since they are the most obvious and readily available biomass. Other plant sources have also been suggested for this purpose. Wood is already important in providing fuel and cash income to small farmers in the developing countries. The areas of research that I will suggest that might be useful for increasing crop yields would probably also apply to increasing the yield of woody species.

As the world population continues to increase and food needs rise further, there will undoubtedly be a greater competition for land and other scarce resources between the requirements for food production and those for biomass formation for energy purposes. Thus, substantial increases in plant productivity per unit of land per unit of time will be necessary. Large improvements will therefore have to be made in the plants themselves and in the efficiency

101

of the processes they carry out during their life cycle that lead
to food and biomass production (Zelitch, 1979a). Imaginative basic
research will be required as well as the ability to translate new
knowledge rapidly into improved crops by methods that will not require substantially greater amounts of labor, fuel, machinery,
fertilizers, or pesticides.

BASIC RESEARCH NEEDS

Basic research is usually defined as being concerned with the
development of knowledge with no predetermined use necessarily in
mind. But actually scientists have some reason for their choice of
problem and the direction the work takes. The choice is probably
not made on a whim merely because the subject exists, but rather
because the scientist believes it may have some importance in that
discipline and that it will ultimately have a widespread ripple
effect. Perhaps I can illustrate this with a personal example. When
I began to investigate the mechanism of action of the enzyme glycolate oxidase in Dr. Severo Ochoa's laboratory early in my career,
it was known that glycolic acid, the substrate of the reaction, was
an early product of photosynthesis and that there was an enzyme in
leaves that rapidly respired this compound. Thus, we decided to
study the biochemistry of this reaction because it might be concerned
with an important respiratory system in leaves. It took a number of
years before I realized how important this reaction was, and that
it was directly involved in the process we now call photorespiration.
This process and its regulation will be discussed later.

Using similar reasoning, one can identify a number of areas
where basic research could provide new knowledge that is needed to
increase biomass production. For example, understanding the control of flowering, fruit development, and other reproductive processes would undoubtedly be helpful. Studies on the regulation of
the translocation of photosynthetic products are needed because the
presence of some metabolites affects net photosynthesis, and translocation provides needed carbon to other plant parts. A knowledge
of methods of decreasing the inhibitory effects of oxygen in the
atmosphere on net photosynthesis in leaves would be most valuable.
The exploitation of the techniques of somatic cell genetics and
molecular biology to effect the transfer of genes between species
and organelles is needed. The union of plant physiology and biochemistry with genetics must play a particularly important role
in providing the tools for solving some of the difficult problems
concerned with increasing crop yields. Specific biochemical
mutants will provide opportunities for testing some hypotheses,
and mutants themselves may untimately provide uniquely valuable
material that will enable plant breeders to obtain large increases
in crop yields.

RELATION OF NET PHOTOSYNTHESIS AND PHOTORESPIRATION TO CROP YIELD

When the CO_2 concentration is increased in a closed system from the normal 0.03% to 0.10%, large increases in net photosynthesis and yield are obtained in many crop species (Hardy and Havelka, 1977). With leguminous crop species they showed that increasing photosynthetic CO_2 fixation also greatly increased symbiotic nitrogen fixation. Clearly raising the CO_2 concentration is not a practical way to increase net photosynthesis for plants growing outdoors because the additional CO_2 diffuses away rapidly and virtually none becomes available to the plants. Nevertheless, these experiments do support the view that if net photosynthesis were increased greatly under field conditions at normal CO_2 levels, crop yields would rise sharply. This may be achieved by using chemical or genetic means to decrease the wasteful photorespiration that occurs in many plant species.

A special kind of light respiration, called photorespiration, takes place in illuminated green tissues of many species. The recognition of the importance of photorespiration was achieved only after a number of obstacles associated with its detection and assay were overcome, and some aspects of the history of this subject have been presented (Zelitch, 1971, 1979a, 1979b). It is now clear that the biochemical reactions responsible for photorespiration differ from those of the more common "dark" respiration. Also the rate of photorespiration is three to five times greater than dark respiration in all C_3 species (Table I). Photorespiration results from the oxidation of compounds produced during photosynthesis, and by blocking photorespiration, net photosynthetic CO_2 incorporation in many species is increased by at least 50%. Table I shows that by a number of different methods of assay, all of which underestimate the rate to some extent, the release of CO_2 by this process consumes about 50% of the net CO_2 assimilated in C_3 species while it is barely detectable in C_4 species.

The C_4 crop species such as maize, sugarcane, and sorghum have low rates of photorespiration and hence have rates of net photosynthesis about twice as great as C_3 species. Average yields and average seasonal growth rates (dry weight produced per square meter of land area per week) are two or three times higher for C_4 species than C_3 species such as wheat, soybean, tobacco and hay.

THE GLYCOLATE PATHWAY OF PHOTORESPIRATION AND ITS REGULATION

It became apparent in the early 1960's that the release of photorespiratory CO_2 had characteristics almost precisely identical to those that controlled the rate of synthesis and metabolism of glycolic acid in leaves (Zelitch, 1979a). Later work has confirmed

Table I.   Minimal rates of photorespiration in various species
           found in published reports (from Zelitch, 1975a, 1979b).
           The assays were performed by a number of different
           methods at 25° to 35°C. Net photosynthesis in leaves was
           determined in normal air at high light intensities. C_3
           species produce phosphoglyceric acid (a C_3 compound) as
           the first product of photosynthesis, while C_4 species
           synthesize oxaloacetate (a C_4 compound) as the first
           product.

Species	Photorespiration rate, % of net photosynthesis
C_3 Plants	
Alfalfa	36
Potato	50
Soybean	42–75
Sugar beet	34–55
Sunflower	27–31
Tall fescue	36–47
Tobacco	25–45
Wheat	17–69
C_4 Plants	
Maize	0–6

this view, and the evidence is presented elsewhere (Zelitch, 1971,
1975a, 1979a). For glycolate to be metabolized, it first must be
oxidized to glyoxylate by the reaction catalyzed by the enzyme
glycolate oxidase. Long-term blocking of this reaction would not
seem a practical way to regulate photorespiration, because glycolate
would accumulate and probably reach toxic concentration. Chemical
analogs of glycolate, alpha-hydroxysulfonates, have been useful
biochemical inhibitors of this reaction, however, in demonstrating
the importance of the glycolate pathway in photorespiration (Zelitch,
1971).

     Several years ago, I began a search for a biochemical inhibitor
of glycolate synthesis that would specifically block photorespiration and increase CO_2 assimilation in leaf disks of a C_3 species,
tobacco. In this assay, leaf disks are cut with a sharp punch and
floated on water or solutions of the biochemical inhibitor being
tested. Glycidate (2,3-epoxypropionate) was found to inhibit glycolate synthesis and photorespiration about 50%, and it increased

net photosynthetic CO_2 assimilation about 50% (Zelitch, 1974).
Glycidate also inhibits glycolate synthesis in maize leaf disks,
but net photosynthesis is not affected, probably because the rates
of glycolate synthesis (about 10% of that for C_3 species) are
already so low in maize that altering glycolate formation has
little effect on net CO_2 uptake. The low rates of glycolate
synthesis in C_4 species undoubtedly account for their slow photorespiration and greater net photosynthesis.

The products of $^{14}CO_2$ assimilation in the presence and absence
of glycidate were examined in tobacco leaf disks, and, as expected,
if glycolate synthesis were inhibited, the concentrations of
glycine and serine were decreased by the inhibitor. The aspartate
and glutamate concentrations were about twice as great in leaf
disks treated with glycidate, but the significance of this observation was not appreciated until several years later.

The metabolic regulation of photorespiration by inhibition of
glycolate synthesis (or any other step of the glycolate pathway)
had not been demonstrated until recently. To assist in laying a
biochemical foundation for genetically altering photorespiration
(Zelitch, 1975b; Berlyn and Zelitch, 1975; Berlyn, 1978), Dr.
David J. Oliver, one of my colleagues, began investigating whether
floating leaf disks on solutions of common metabolites would
inhibit glycolate biosynthesis. The assay used previously to detect
the inhibitory effect of glycidate was employed. These experiments
revealed that increasing the intracellular concentration of aspartate, glutamate, phosphoenolpyruvate, or glyoxylate by floating
leaf disks on solutions of these compounds in the light causes
glycolate synthesis and photorespiration to be inhibited (Oliver
and Zelitch, 1977a, 1977b). The $^{14}CO_2$ uptake in the light was
increased as much as two-fold in the presence of glyoxylate (Fig.
1). Thus, modest alterations in the concentrations of several
common metabolites can increase photosynthesis by blocking
glycolate synthesis and photorespiration, probably by a feedback
type mechanism.

Another colleague, Dr. Arthur L. Lawyer, later found that
glycidate strongly inhibits a reaction of the glycolate pathway,
the glutamate:glyoxylate aminotransferase reaction (Lawyer and
Zelitch, 1978). The effect of glycidate is thus probably an indirect one brought about by increases in the pool sizes of glutamate, aspartate, and glyoxylate when the aminotransferase reaction
is blocked.

Floating leaf disks on glyoxylate solution in the light also
brings about a decrease in the inhibition of net photosynthesis by
oxygen (Oliver, 1978). When the O_2 concentration was reduced from
21% to 3%, net photosynthesis was increased 60%, while it only

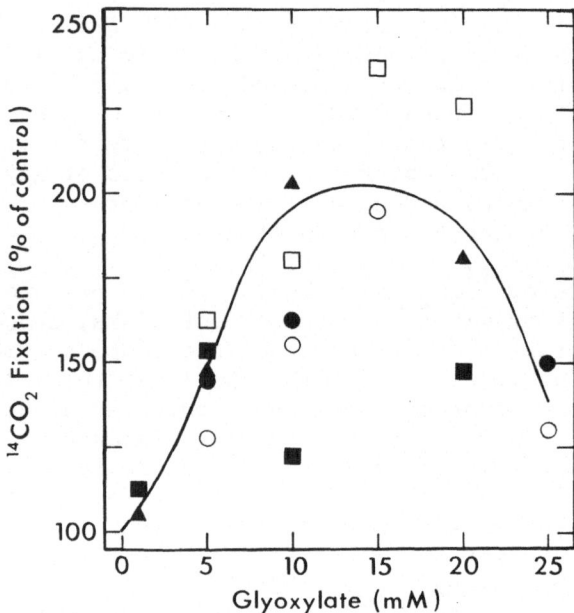

Fig. 1.   Effect of glyoxylate concentration on increase in net
          photosynthetic CO_2 fixation in tobacco leaf disks (from
          Oliver and Zelitch, 1977b). The leaf disks were floated
          in the light on glyoxylate solutions for 1 hr before the
          photosynthetic measurements were made.

increased 22% in the presence of glyoxylate (Table II). The
results obtained on the regulation of glycolate synthesis and
photorespiration in leaf tissue provide evidence that chemical or
genetic regulation of the formation and metabolism of some commonly
occurring compounds should produce plants with decreased photorespiration and higher rates of net photosynthesis.

PRODUCING MUTANTS OF HIGHER PLANTS BY SELECTIONS ON PLANT CELLS

      Several colleagues and I are attempting to develop and analyze
a mutational approach to the regulation of photorespiration using
plant cells. In this approach cell and callus cultures of tobacco
have been used as a model system in attempts to mimic the production
of biochemical mutants that has been so successful in bacterial
and fungal systems. The tobacco system allows regeneration of
plants and analysis of the genetic nature of any mutants obtained
(Berlyn, 1978; Zelitch, 1979a).

      Dr. Mary Berlyn has produced cell cultures resistant to several
inhibitors that affect the glycolate pathway. An example of one

Table II. Effect of glyoxylate on the stimulation of net photo-
           synthesis by lowered O_2 concentrations in the atmos-
           phere (from Oliver, 1978). Net photosynthesis was
           determined on tobacco leaf disks previously floated on
           15 mM glyoxylate solution for 1 hr in the light.

Treatment	Rate of Net Photosynthesis		Stimulation by Lowering O_2 to 3%
	21% O_2	3% O_2	
	μmol $^{14}CO_2$/g fresh wt$\cdot$hr		%
Water	59.1	94.7	60
Glyoxylate	82.0	99.7	22

study she has conducted utilizes a strategy of altering the metabolism of glycine and serine, since these compounds might act as
regulators of the photorespiration pathway. She has described
(Berlyn, 1978) the isolation of 20 cell lines that when grown on a
sucrose medium are stably resistant to isonicotinic acid hydrazide
(INH), an inhibitor of the conversion of glycine to serine. A
number of plants were produced from 13 of the variant INH-resistant
cell lines, and tests on callus from these plants, and, in some
cases from seedlings of the next generation, indicate that for at
least some lines the INH-resistant characteristic has been retained
and transmitted (Berlyn, M.B., manuscript in preparation). Thus,
biochemical mutants were produced.

We are currently studying the effects of INH on glycolate
metabolism in cell cultures. As expected, a dramatic effect is the
blocking of the conversion of glycine to serine by INH in normal
cells. The variant cell lines have been examined for changes in
the sensitivity to INH inhibition of this conversion. We showed
that the resistance of these cells in INH is not caused by their
inability to take up the inhibitor or to metabolize INH more
rapidly than susceptible cells.

At least 5 of the INH-resistant cell lines show greater resistance than wild type cells to INH inhibition of the decarboxylation
of glycine to produce CO_2 and serine in particulate preparations
isolated from the cells (Table III). In preparations obtained from
two of the resistant cell lines that were studied more extensively,
the K_i for INH inhibition, which is inversely related to the binding
of INH, was two- to four-fold greater than that of wild type preparations. Plants are available from seed of some of these INH-resistant lines, and it should soon be possible to determine whether
these changes in biochemical phenotype produced by selection on

Table III.   Resistance of glycine decarboxylation activity to INH
             in particulate preparations from INH-resistant tobacco
             callus (from Berlyn, M.B., and Zelitch, I., unpublished).
             Inhibition in wild type preparations averaged 56%.
             Numbers of preparations assayed are shown in parentheses.

Cell line	Average difference in % inhibition from wild type in presence of 0.25 mM INH
UINH 5	−18 (4)
UINH 18	−13 (10)
UINH 21	−23 (3)
UINH 24	−15 (7)
UINH 40	−10 (4)

callus are expressed in the whole plant.

We have also grown tobacco callus cultures through many pass-
ages in the light with an elevated CO_2 atmosphere (1 to 3% CO_2) as
the only carbon source (Berlyn, et al., 1978). This photosynthesis-
dependent growth can continue indefinitely, and these kinds of cells
should permit the direct selection of mutants with altered photosynthetic properties.

As in C_3 plants, these autotrophically grown cells were advers-
ly affected by high O_2 levels because of the increase in photorespiration and the corresponding decrease in net photosynthesis.
Thus, 60% O_2 proved to be toxic during three-week incubations to
callus cells grown on CO_2 but not to cells grown on sucrose (Berlyn,
et al., 1978). Almost all the cells were killed by prolonged
exposure to high O_2 levels, but several cell lines were recovered
by later incubation on a sucrose medium.

These results indicate it may be feasible to select lines for
O_2-resistance, and perhaps some of these cell lines will be resistant because of a lowered rate of photorespiration. It thus appears
that the areas of photosynthesis and photorespiration will provide
special opportunities that will lend themselves to exploitation
by the powerful techniques of somatic cell genetics to increase
photosynthetic uptake of CO_2 and increase crop yields.

ECONOMIC BENEFITS OF AGRICULTURAL RESEARCH

A number of economists have analyzed the annual return of investment in agricultural research in developed and developing

countries. Some of these results are summarized in Table IV, and
they show that the returns are considerably higher than those realized in common business investments. The returns are high in all
crops and all countries.

These economic investigations prove the importance of science,
and that building and maintaining research capability also offers
developing countries a unique opportunity to purchase a tremendous
amount of growth (Evenson, 1974). There is a time lag between investment and benefits, however, so that research done now will
typically require a period of 10 years before increases in productivity are realized. Hence research requires stability, both
financial and political, and long term commitments.

Basic research when taken seriously is a lonely, difficult and
often frustrating undertaking, and it is particularly sensitive to
unpredictability in funding. It is a necessity and not a luxury,
even in developing countries, and experience shows it will be most
effective if institutions are created where theory and practice are
carried out together. The failure to appreciate the necessity of
uniting basic agricultural research with practical was illustrated
recently by an advisor to the World Bank who saw basic research in
developing countries as merely an inexpensive way to keep university
faculty from emigrating and mainly useful for the training of graduate students (Weiss, 1979).

Farmers and their legislative representatives must be convinced
of the wisdom of investing in science even though the benefits are
often unknown and cannot honestly be described in advance with any
precision. Similarly, scientists must be provided with the capability and must have the courage to test their laboratory results
in the field.

Table IV. Estimates of the rates of return from investment in
           agricultural research (from Wortman and Cummings, 1978).

Country	Commodity	Period	Annual Return, %
USA	Hybrid maize	1940–55	35–40
USA	All Research	1949–59	47
Mexico	Wheat	1943–63	90
Mexico	Maize	1943–63	35
Bolivia	Rice	1957–64	79–96
Columbia	Rice	1957–72	60–82
Japan	Rice	1930–61	73–75
India	All Research	1953–71	40

THE ORGANIZATION OF SCIENCE FOR MAXIMUM EFFECTIVENESS

My views about the organization of science are obviously influenced by my personal experiences. All of my professional career, I
have been doing fulltime research on basic problems related to plant
productivity at the Connecticut Agricultural Experiment Station.
The U.S. Department of Agriculture supports a great deal of agricultural research, and there are Agricultural Experiment Stations in
each of the 50 states. Ours is one of the smallest, having about
45 scientists working in seven departments related almost entirely
to the plant sciences. Virtually all of my time is spent working
in the laboratory on experiments carried out by myself in collaboration with several colleagues. Except for a grant of short duration
from the Rockefeller Foundation, which was most helpful at a critical time, our work has been supported directly and entirely by
Experiment Station funds (as described below) without the aid of
extramural research grants. I have served terms on grant evaluation
panels for both the National Science Foundation and the National
Institutes of Health, so I am familiar with the procedures and
policies of the competitive grants system.

Before World War II, agricultural science in the public sector
of the U.S. was financed almost entirely by the U.S. Department of
Agriculture and the State Agricultural Experiment Stations. The
Hatch Act of 1887 provided direct grants from the federal government
to the states to support the Experiment Stations without providing
any overhead costs, with matching expenditures required by the
Stations, and this policy of federal support has continued to the
present. About 21 percent of Experiment Station funds now come from
these so-called federal formula funds, and about 68 percent from
state appropriations.

In 1875, Samuel W. Johnson, a professor of Agricultural Chemistry at Yale, lobbied the Connecticut legislature into establishing
the first American Experiment Station. He clearly conceived that
the "principles and applications" of agricultural sciences should
be studied side by side. Thus he saw the importance of basic
research in the Experiment Station system, and this principle was
followed in the Stations established soon afterwards in all the
states. Evenson (1978) has calculated the increases in the productivity of American agriculture from 1868 to 1926 related to the
research performed by the state Experiment Stations and the U.S.
Department of Agriculture; the return was 65 percent annually. The
effectiveness of this research system might provide a model for the
organization of research in the developing countries.

After World War II, agricultural science in the U.S. was
financed increasingly outside the Department of Agriculture. The
increased funding for research in Agriculture was small in comparison, for example, with the increase in funds for the National

Table V. The compound percentage increase of federal funds for
          research and development (from Evenson, Waggoner, and
          Ruttan, 1979). Amounts were reduced to 1967 dollars by
          the consumer price index, and the compound rates of
          changes were calculated from 1975 amounts divided by 1955
          or 1969 amounts.

Period	Agriculture	National Science Foundation
1955–75	5.4	18.4
1969–75	1.6	6.1

Science Foundation. This small increase in Agriculture occured
even from 1969 to 1975, in spite of rapid growth in agricultural
exports and world food crises in the mid-1960's and early 1970's
(Table V).

      The National Science Foundation funding policies are greatly
different from the Experiment Station system. It is almost exclu-
sively for basic research, and it largely supports scientists at
universities who work in a single discipline. Some university
scientists have come to believe that they should do the basic
science and the Experiment Stations should stick with the practical.
In their interesting book on feeding the world, Wortman and Cummings
(1978) state that the National Science Foundation reportedly did
not fund many studies of economically important species on the
grounds that they are "applied" and therefore the responsibility of
the U.S. Department of Agriculture. On the other hand, the U.S.
Department of Agriculture reportedly did not support basic research
because they considered it the primary responsibility of the
National Science Foundation. This separated basic from practical
research, and coupled with the decrease in funding for agricultural
research, it has diminished the research effectiveness in the U.S.
for a number of years.

      I acknowledge that the National Science Foundation has funded
important and excellent research that will be useful in increasing
future crop productivity. I am concerned, however, that the
policy of separating out and only funding basic research will
increase the time needed for making basic discoveries useful.

      Let us consider some specific difficulties. The competitive
grants system at present does not generally provide funding for
periods greater than 3 years. Many kinds of research require a
longer commitment, and experienced professional scientists cannot
usually be hired for short periods only. About one to two months

is consumed in preparing grant proposals, and this is a large fraction of a scientist's time that could be better spent in the library, at the bench, or in the field. The peer review system takes up considerable time, and though it separates good from bad research, highly unorthodox yet important ideas will often not be accepted by peers as the history of science shows repeatedly. Peer groups are reluctant to cut off unproductive researchers requesting grant renewals, a characteristic I observed often while serving on panels. In addition to the time spent by panel members reviewing grants and visiting sites, large amounts of time are donated by a vast network of reviewers who assess proposals by mail. Last year, for example, I alone reviewed 20 proposals by mail for the National Science Foundation. If my experience is a common one, the time consumed in this activity nationwide must be immense.

The growth of the present National Science Foundation grants system has had some other adverse effects on science not often appreciated by those dependent on it. The entrepreneurial aspects of obtaining grants often cause young scientists to lose sight of what research is all about. When I visit universities, for example, I cannot tell questioners how many square feet my laboratory comprises, a statistic every grant-seeking investigator apparently knows by heart. At some universities the total dollar amount of a scientist's grants is used as an important criterion for promotion. Above all, the grants system does not generally lend itself to funding proposals by small interdisciplinary teams, something that is especially needed for increasing the efficiency of biomass production. In the past when interdisciplinary work was encouraged by some programs in the National Science Foundation, the funding times were short and the filing of progress reports burdensome. Since the new U.S. Department of Agriculture Competitive Grants program has copied the National Science Foundation system, I think it is fair to say that it, too, will probably suffer from the same disadvantages.

The strength of agriculture in this country and the strength of its scientific base is due to our having a pluralistic system of supporting this science. But the Experiment Stations play an essential role within this system because they are specifically designed to have both laboratories and field plots. They must not become mediocre by reducing their funding.

The field of photosynthesis demands that people in several disciplines cooperate if crop productivity is to be increased quickly. Scientists need to work on enzymes, organelles, leaves and crops in disciplines of biochemistry, plant physiology, molecular biology, genetics, plant breeding, and agronomy. We must have competent people to work at all these levels of biological organization and in all these subjects.

One of the advantages of the Agricultural Experiment Station system is that a Station is present in each State and they are therefore able to react quickly to local problems. The combining of basic research with that oriented to practice undoubtedly came about not because scientists wanted it but because citizens did. The system provides an opportunity for scientists in all the disciplines mentioned above to serve either under the same roof or in close proximity and to solve short-term and long-term problems.

Organization by itself is no guarantee of excellence, but scientists do not work in a vacuum, and an organization is needed that facilitates cooperation between scientists working at different levels of biological organization and from different disciplines. Ideally, scientists must have an opportunity for doing research unencumbered insofar as possible from lengthy proposal writing, lengthy preparation of internal reports, and uncertainties and unpredictable funding. The Experiment Station system is not the only way to accomplish this, but it has been in existence in this country for over 100 years, and it has worked to increase agricultural productivity as Table IV shows. This system should be encouraged and used as a model for other groups, especially in developing countries, who must cooperate in both practical and laboratory science. Unless theory and practice are united, the food supply will be loser, as will all of us.

Acknowledgements: Helpful discussions with my colleagues Mary B. Berlyn, Kenneth R. Hanson, James G. Horsfall, David J. Oliver and Paul E. Waggoner are gratefully acknowledged.

REFERENCES

Berlyn,M.B., 1978, A mutational approach to the study of photo-respiration, in: "Photosynthetic Carbon Assimilation," H.W. Siegelman and G. Hind, eds., Plenum, New York.
Berlyn, M.B., Zelitch, I., and Beaudette, P.D., 1978, Photosynthetic characteristics of photoautotrophically grown tobacco callus cells, Plant Physiol., 61: 606-610.
Berlyn, M.B., and Zelitch, I., 1975, Photoautotrophic growth and photosynthesis in tobacco callus cells, Plant Physiol., 56: 752-756.
Evenson, R.E., 1974, Science and the world food problem, Conn. Agr. Expt. Sta. Bull. 758.
Evenson, R.E., 1978, Center discussion paper 296 (Economic Growth Center, Yale University, New Haven).
Evenson, R.E., Waggoner, P.E., and Ruttan, V.W., 1979, Economic benefit from research: An example from agriculture, submitted for publication.

Hardy, R.W.F., and Havelka, U.D., 1977, Possible routes to increase
    the conversion of solar energy to food and feed by green
    legumes and cereal grains in: "Biological Solar Energy Conver-
    sion," Mitsui, A., et al., eds., Academic Press, New York.
Lawyer, A.L., and Zelitch, I., 1978, Inhibition of glutamate:
    glyoxylate aminotransferase activity in tobacco leaves and
    callus by glycidate, an inhibitor of photorespiration, Plant
    Physiol., 61: 242-247.
Oliver, D.J., 1978, Effect of glyoxylate on the sensitivity of net
    photosynthesis to oxygen (the Warburg effect) in tobacco,
    Plant Physiol., 62: 938-940.
Oliver, D.J., and Zelitch, I., 1977a, Metabolic regulation of
    glycolate synthesis, photorespiration, and net photosynthesis
    in tobacco by L-glutamate, Plant Physiol., 59: 688-694.
Oliver, D.J., and Zelitch, I., 1977b, Increasing photosynthesis
    inhibiting photorespiration with glyoxylate, Science 196: 1450-
    1451.
Weiss, C., Jr., 1979, Mobilizing technology for developing countries,
    Science 203: 1083-1089.
Wortman, W., and Cummings, R.W., Jr., 1978, "To Feed this World.
    The Challenge and the Stretegy," Johns Hopkins Univ., Baltimore.
Zelitch, I., 1971, "Photosynthesis, Photorespiration, and Plant
    Productivity," Academic Press, New York.
Zelitch, I., 1974, The effect of glycidate, an inhibitor of glyco-
    late synthesis, on photorespiration and net photosynthesis,
    Arch. Biochem. Biophys., 163: 367-377.
Zelitch, I., 1975a, Pathways of carbon fixation in green plants,
    Ann. Rev. Biochem., 44: 123-145.
Zelitch, I., 1975b, Improving the efficiency of photosynthesis,
    Science 188: 626-633.
Zelitch, I., 1979a, Photosynthesis and plant productivity, Chemical
    and Engineering News 57: 28-48.
Zelitch, I., 1979b, Photorespiration: studies with whole tissues,
    in: "Encyclopedia of Plant Physiology, Vol 6," M. Gibbs and
    E. Latzko, eds., Springer-Verlag, Berlin.

BIOLOGICAL NITROGEN FIXATION: A FERTILIZER STRATEGY POTENTIALLY

BENEFICIAL FOR THE POOR IN DEVELOPING COUNTRIES

Marvin R. Lamborg

Charles F. Kettering Research Laboratory
150 East South College Street
Yellow Springs, OH   45378   USA

Food is the world's most important renewable resource and plants
constitute directly or indirectly (as feed) essentially all of the
world's food. In publishing "The Limits of Growth", the Club of Rome
voiced concern not only for the continuing problems of hunger, malnutrition, disease and death, but also for a perceived inability to
keep pace with food needs in the future (Meadows et al., 1972).
After the World Food Conference in Rome in 1974, President Gerald
R. Ford requested of the U.S. National Academy of Sciences that it
"make an assessment of this problem and develop specific recommendations on how our research and development capabilities can best
be applied to meeting this major challenge". In 1977 the Academy
responded with an in depth report "World Food and Nutrition Study,
The Potential Contributions of Research" pointing to the need for
a sustained 3 to 4% annual growth rate of food production in
developing countries.

Today the yield of rice in southeast Asia shows signs of level-
ing off and there are similar prospects for maize, potatoes, wheat
and cassava in Latin America. Overall, grain yields in the world
are in decline. By way of contrast, India has enjoyed
a significant surplus of grains (especially wheat) in the last 3
years. The paradox is that the poorest farmers have not shared in
the benefits of the bountiful harvest because they didn't have the
technology, fertilizer and water required by the miracle grains; so
they continue subsistence farming. Of the commonly required fertilizer constituents nitrogen is the most expensive. The poorest
farmers apply little fertilizer N and generally do not plant crop
varieties which are nitrogen responsive. What is needed in the

Contribution No. 653 from C.F. Kettering Research Laboratory

115

democratic developing countries is appropriate technology which will
boost yield and purchasing power of the world's poorest farmers.
Biological nitrogen fixation is just such a technology.

This report highlights examples of new knowledge and understanding of nitrogen-fixing symbiotic associations, and the relationship between the processes of nitrogen fixation and photosynthesis and of the iron-molybdenum cofactor of nitrogenase. These
are areas of biological nitrogen fixation likely to prove useful for
the poorest farmers in developing countries when the new information
is applied in a useful way. A thorough treatment of the subject
has not been made in this report. Rather, my purpose is to show
that a good deal of new information and insight has been gained in
the last 2 to 3 years and that there is forward movement across a
broad front of science disciplines.

NITROGEN FIXATION BY PURE CULTURES OF RHIZOBIUM

Until recently nitrogen fixation research in legumes was
hampered by the necessity of infecting the host plant with the
appropriate microorganism in order to elicit nitrogenase activity.
In fact, because nodule ontogeny was a prerequisite for the synthesis of nitrogenase, it was difficult to construct experiments
capable of pinpointing a cause for the lack of nitrogenase activity
because of the interplay of nodule development and nitrogenase
synthesis. In 1975 five different laboratories reported nitrogen
fixation in pure cultures of rhizobia (Keister; Kurz and LaRue;
McCombs et al.; Pagen et al.; and Tjepkema and Evans). The key
to derepression, where it is successful, is to reduce the partial
pressure of oxygen to less than 0.5%. Figure 1 depicts the rate of
acetylene reduction as a function of pO_2. The optimum pO_2 concentration is 0.13%, even at higher concentrations acetylene reduction
is observed after a lag period. In 1975 only a few strains of R.
japonicum and the cowpea species could be derepressed in pure culture. Now at least half of the R. japonicum strains tested are
capable of reducing acetylene although the rates of reduction vary
considerably from strain-to-strain (Keister, personal communication).

Some strains of R. japonicum which remain repressed under
reduced oxygen tension can be derepressed by a factor(s) derived
from the media in which soybean cells have been grown in suspension
culture (Reporter, 1978). The factor(s) has been fractionated, and
partially characterized (Storey and Reporter, 1978; Skotnicki et al.,
1978). Most of the reports of nitrogen fixation with Rhizobium have
used strains of the slow-growing species. For some reason the fast-
growing species seem much more refractory to derepression. It
would be desirable to derepress the fast growers because most of
the genetic analysis has been reported using them. Skotnicki et al.
(1979) have recently reported that pure cultures of a few

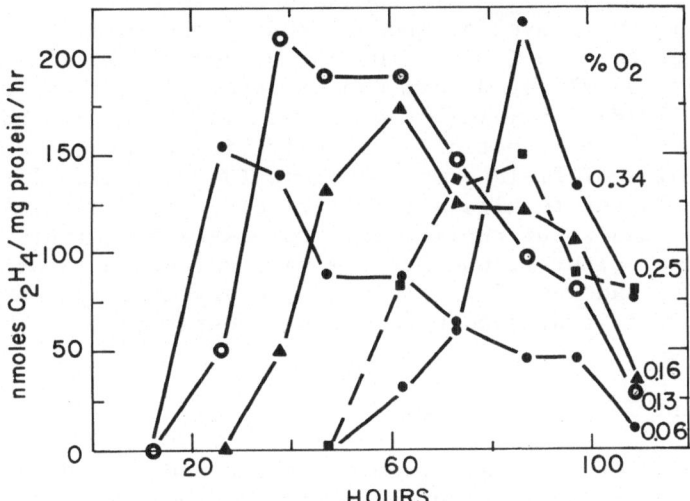

D.L. Keister and V. Ranga Rao (1976) Academic Press, New York, p. 419-430

Fig. 1.   Rates of acetylene reduction by R. japonicum 61A76 as a
function of the partial pressure of oxygen.

Spectinomycin resistant (spontaneous and induced) mutants are also
able to reduce acetylene under conditions where the original parent
culture was refractory. If the observation proves to have general
applicability it will be most exciting and important.

GLUTAMINE SYNTHETASE IN RHIZOBIUM

      In microorganisms, glutamine synthetase (GS) has a bipartite func-
tion. On the one hand, the protein plays a catalytic role in the
utilization of ammonia and the production of glutamine. Additionally,
it regulates the synthesis of numerous enzymes involved in nitrogen
metabolism (Ginsburg and Stadtman, 1973; Wohlhueter et al., 1973;
Magasanik et al., 1974). In most bacteria a single protein is responsible for carrying out both functions; as a positive regulator
of transcription it is thought to be primarily deadenylylated; the
enzymically active-form is also deadenylylated. Rhizobia grown
aerobically in the absence of ammonia have two forms of glutamine
synthetase. One form appears to have the same bipartite function

previously cited. The novel form is not adenylylated (Darrow and
Knotts, 1977). In contrast, enteric bacteria and the classical
free-living nitrogen fixers A. vinelandii and K. pneumoniae have
only one GS. Some of the physical characteristics of the two
proteins from R. japonicum are compared in Table 1. During aerobic
growth, GS II constitutes as much as 80% of the total GS. Under
reduced oxygen tension (conditions for nitrogen fixation) only GS I
in the highly adenylylated form can be detected as shown in Figure 2
(Rao et al., 1978). Similarly, only GS I in the highly adenylylated form has been detected in actively fixing bacteroids isolated
from the nodules of field-grown soybeans (Darrow, unpublished results). The latter results are preliminary, based on limited
sampling and need to be replicated. How oxygen regulates GS synthesis in pure cultures, whether the same mechanism is operative
in bacteroids, and why GS I is highly adenylylated when nitrogenase
is being actively expressed needs to be understood.

EXPERIMENTALLY INDUCED BACTEROIDS?

When the microorganisms from a soybean nodule homogenate are
fractionated by a stepwise sucrose density gradient, three distinct
bands are obtained. These are termed mature bacteroids, transforming bacteria and bacteria (Ching et al., 1977). Figure 3 shows
the distribution of microorganisms into the several classes cited in
this paper. Thus a soybean nodule is a heterogenous collection of
microorganisms. Evans and Crist (unpublished) have carried out a
similar analysis using nitrogen-fixing pure cultures of cowpea
rhizobia 32H1 (Fig. 4). At zero time there is essentially a single
species of microorganism. With time the population becomes more
heterogeneous unless ammonium is present. In this study 66% of the

Table 1.   Properties of GSI and GSII

GSI (Conventional)	GSII (Novel)
Heat Stable (50°)	Heat Labile (50°)
Apparent pI = 5.4	Apparent pI = 6.1
S_{20w} = 20S	S_{20w} = 11.5S
Undergoes Adenylylation	No Adenylylation
EM = Dodecamer with Stacking	EM = Hexamer?, No Stacking
M = 706,800 (12 x Subunit)	$M_w \cong$ 200,000? (High Speed Equilibrium)
$K_m NH_4^+ < 0.2mM$	$K_m NH_4^+ < 0.2mM$

Darrow and Knotts, unpublished.

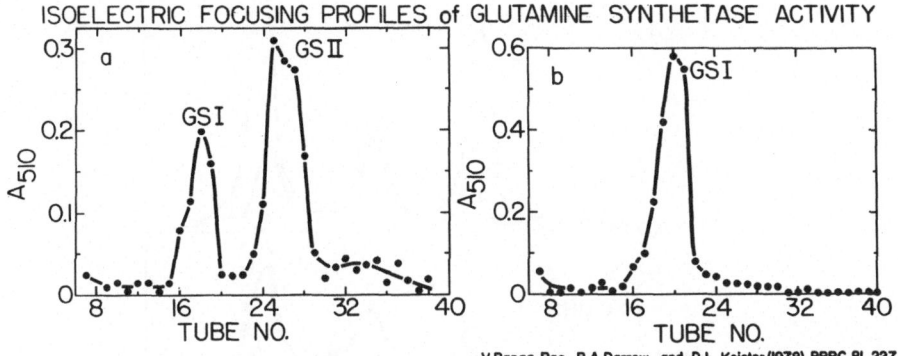

ISOELECTRIC FOCUSING PROFILES of GLUTAMINE SYNTHETASE ACTIVITY

V. Ranga Rao, R.A. Darrow, and D.L. Keister (1978) BBRC 84, 227

Fig. 2.   Isoelectric focusing profiles of glutamine synthetase in
cell-free extracts of <u>R</u>. <u>japonicum</u> 61A76.   Cells are grown
at 0.75% (panel a) and 0.16% (panel b) average O_2 concentration.

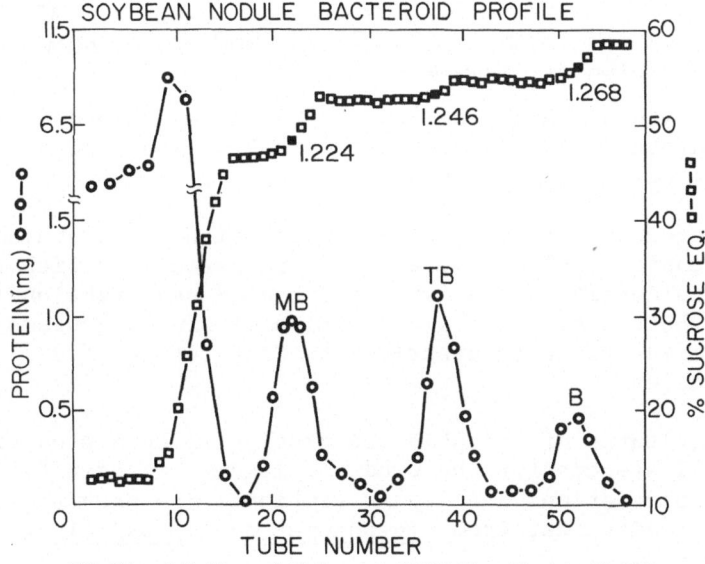

SOYBEAN NODULE BACTEROID PROFILE

T.M. Ching, S. Hedtke and W. Newcomb (1977) Plant Physiol. 60, 772

Fig. 3.   Stepwise sucrose density gradient centrifugation of the
postnuclei homogenate from soybean nodules.   MB: mature
bacteroids, TB: transforming bacteria, B: bacteria.

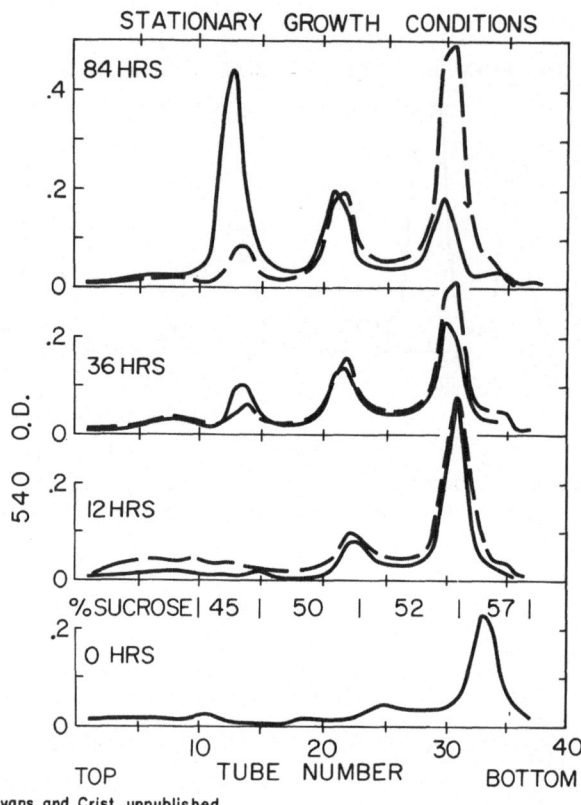

Fig. 4.   Growth of <u>Rhizobium</u> cowpea 32H1 on a defined minimal salts,
          carbon media plus ammonium (dashed line) and without
          ammonium (solid line)

total nitrogenase activity is associated with the organisms of
lightest density, 28% of the activity is associated with those of
intermediate density and 6% with the most dense fraction (probably
unmodified bacteria). The distribution of microorganisms shown
here is similar to those described by Ching et al. (1977) from
soybean nodules.

     Pure cultures of rhizobia are proving to be a good experi-
mental model systems for the study of legume nitrogen-fixing
symbiotic associations and legume symbionts are proving to be
sufficiently different from <u>Azotobacter</u> and <u>Klebsiella</u> to warrant
special attention.

     The free-living rhizobia model system permits the researcher
to dissect out the nitrogen fixation process from the complex
process of root nodule ontogeny thus greatly simplifying studies
of the genetics and control mechanisms in these symbiotic microorganisms.

ORGANIC NITROGENOUS COMPOUNDS SYNTHESIZED FOR TRANSPORT
AND ASSIMILATION

Nitrogen is fixed by bacteroids in soybean nodules; where is
ammonia converted to an organic form and assimilated? Klucas and
Burris (1967) showed that almost all the N appears in the non-
bacteroid portion of the nodule and Bergersen and Turner (1967)
using isolated bacteroids found that 95% of the N_2 fixed is
excreted as ammonia. More recently, other workers, using pure
cultures of nitrogen fixing rhizobia verified the excretion of
ammonia (O'Hara and Shaumugan, 1976; Tubbs,1976; Bergersen and
Turner, 1978; Upchurch and Elkan, 1978). So it is the plant cells
which convert ammonia to an organic product for transport and
assimilation. In the early seventies glutamine and asparagine
were identified as major N constituents of cell sap. Mothes (1961)
pointed out that ureides could also be a major nitrogen constituent
of plant cells (including legume plants). More recently Yamamoto
and coworkers (Fujihara et al., 1977; Matsumoto et al., 1977a,b,c)
have demonstrated that the primary nitrogenous compound found in
the stem exudate of soybeans is allantoin. Streeter has confirmed
and extended those observations by showing that allantoin and
allantoic acid are the predominant nitrogenous compounds transported
from nodules during reproductive growth. Prior to the time of reproductive growth, amino acids, asparagine and ureide N are excreted
in roughly equal amounts. Allantoin is the principal nitrogenous
compound excreted from cowpea nodules as well (Herridge and Pate,
1977). It should be noted that allantoin is not the major transport
compound in all legumes. Asparagine is thought to have the major
transport function in lupine (Pate et al., 1977) and pea (Lewis and
Pate, 1973) and in the non-legume nitrogen-fixing alders it is the
amino acid citrulline (Gardner and Leaf, 1960). It is important to
realize that there are significant metabolic points of diversion
among the legumes. If a breeder working on chick peas chooses to
use nitrogen transport compounds as a diagnostic trait, he must
know what that compound is.

THE RECOGNITION/INFECTION PROCESS IN LEGUMES

A number of plant lectin proteins have been isolated, purified
and chemically characterized. Interest in them developed because
they are hemagglutinin factors – they agglutinate and precipitate
red blood cells with some specificity. Until recently, lectins
were well characterized plant proteins in search of a physiological
function. In the case of a few of the legume lectins one function
is likely to involve recognition of the Rhizobium prior to infection.
Host range specificity – the demonstration that soybean lectin binds
to Rhizobium japonicum (the microbe which infects soybeans) – and
not to other species of Rhizobium was first reported by Bohlool and
Schmidt (1974). Evidence from at least two sources suggests that
the plant lectin protein binds to the polysaccharide of the bacterial

capsule in a reaction analogous to the association of antibodies
and antigens (Bhuvaneswari et al., 1977; Dazzo and Hubbell, 1975).
Lectins should be on the surface of roots (really root hairs) where
infection is known to take place. With the exception of the recent
report by Dazzo and Brill (1977) it is very difficult to demonstrate
the presence of lectins on roots. Recently, Bauer et al. (1979)
have observed that the first nodules are formed on that surface of
the primary root bounded by the growing root tip and the portion of
the root where the smallest-emerged root hairs are visible by light
microscopy - a no root hair zone. There is a 3-4 hour lag period
before any cytological changes can be observed. The lag can be
eliminated by a 3-4 hour pretreatment of the sensitive root zone
with autocalved capsular polysaccharide or N-acetyl galactosamine
(which is a hapten of soybean lectin). In some way the capsular
polysaccharide sensitizes the host to the subsequent events of
infection.

The results of these studies suggest related avenues for investigation for: (a) identifying a possible cause for non-infectivity;
(b) modifying host range specificity - competition for inoculated
soybeans by marginally effective indigenous rhizobia is a serious
problem in much of the soybean growing land in the United States;
(c) extending the legume recognition research to pathogenic infection processes.

GENETIC STUDIES

Most of the genetic analysis which has been reported for
Rhizobium has been carried out with the fast growing species.
Linkage maps have been constructed for R. meliloti (Meade and Signer,
1977; Kondorosi et al., 1977) and R. leguminosarum by plasmid
mediated recombination (Beringer et al., 1978). Mutants of both
slow and fast growing rhizobia have been isolated and described.
Many of these studies relate to infectivity and nodule formation
rather than effectivity as nitrogen fixers. In those cases where
effectiveness was assayed in whole plants or nodules, it is difficult
to pinpoint the lesion because abnormal nodule development may
preclude normal, active nitrogenase synthesis.

High molecular weight (> 10^8 daltons) plasmids have been
identified in several Rhizobium species (Nuti et al., 1977). However, since plasmids are found in infective and non-infective
strains, it is not known whether genes for symbiotic association
are plasmid borne. When the tumor producing TI plasmid from Agrobacterium tumefaciens is transferred to Rhizobium trifolii, the ex-
conjugant found is able to induce tumor formation and form effective
nodules (Hooykaas et al., 1977) depending upon the site of infection.
Finally, a plasmid has been used to bring about an altered host
range specificity (Johnston et al., 1978). The transposon, Tn 5,

is transferred from R. leguminosarum to a non-nodulating strain of
R. leguminosarum and three other species of Rhizobium as shown in
Table 2. It can be seen that the ability to nodulate peas can be
transferred with high frequency which suggested that some of the genes
for nodulation are carried by a plasmid. Only the leguminosarum
and phaseoli transconjugant reduced acetylene at normal rates.

RELATIONSHIP BETWEEN PHOTOSYNTHESIS AND NITROGEN FIXATION

     Nitrogen fixation is an energetically costly process. The
latest reported requirement for ATP by nitrogenase in vitro is
4.5 moles/2e change at 25°, pH 8.1 (Watt et al., 1975). This is
of course solely for the conversion of N_2 to ammonia. In whole
cell systems, Andersen et al. (1977) reason that one should include
energy cost estimates for cell metabolism and for protein turnover
(for longer-term nitrogen fixation) as well. Based on measurements
made with mutants of Klebsiella pneumoniae they estimate the ATP
requirement could be as high as 35-40 moles per mole of N_2 fixed.
There is no circumventing the energy burden for the energetics of
nitrogen assimilation via nitrate reduction is of the same order
of magnitude (Hardy and Havelka, 1975). Regarding their potential
use in genetic engineering, Valentine (1977) suggests the limitation
of energy for growth, maintenance and nitrogen fixation is so severe
in nitrogen-fixing (higher plant) systems that they pose no

Table 2. Transfer of Ability to Nodulate Peas and Reduce Acetylene
          in the Nodules from Strain T3 to Other Strains of Rhizobium

Recipient strain	Species	Markers	Control clones No. nodu- lating/ no. tested	Transconjugants clones No. nodu- lating/ no. tested	No. reducing C_2H_2/ no. tested
6015	R.leguminos- arum	phe1 trp12rf + 392 str37inf	0/9	17/17	8/8
6661	R.trifolii	rif	0/7	16/16	2/9
6710	R.trifolii	rif	1/10	17/17	1/8
1233	R.phaseoli	rif	0/6	9/9	9/9
9009	R.'species'		0/10	8/10	2/8

A.W.B. Johnston et al. (1978) Nature 276, 635.

ecological threat to the environment.

Hardy and Havelka (1975) demonstrated that photosynthate limits nitrogen fixation in soybeans under field conditions. In field experiments these investigators reported that as a result of increasing pCO_2 to 1000 ppm or more, total nitrogen increased 1.7-fold. In control plots 26% of the total nitrogen in the plants was derived from nitrogen fixation. As a result of the CO_2 enrichment, 83% of the nitrogen originated from nitrogen fixation. Hardy and Havelka conclude that the major effect of elevated CO_2 is to reduce photorespiration and increase the available photosynthate. Recently, Bethenfalvay et al. (1978a, 1978b) observed that both infection and the nitrogen-fixing efficiency of the infecting microorganism affects the characteristics of growth and photosynthesis. This suggests that the processes of nitrogen fixation and photosynthesis though physically separated in higher plants are in chemical communication and modulate each other. The nature and locus of this regulatory site need to be identified. The amount of nitrogen fixed can also be increased by extending the period of active nitrogen fixation by delaying nodule senescence. Abu-Shakra et al. (1978) have identified varieties of soybean with apparent uncoupling of vegetative and reproductive growth. The leaves remain green throughout the reproductive seed maturation period and high rates of nitrogen fixation are sustained for longer periods of time. Some of their results are presented in Table 3. The F_3 plants 1-3 showed delayed leaf senescence and a positive correlation between key parameters and nitrogen fixation. The other plants (4-6) demonstrated normal senescence.

EFFICIENCY OF NITROGEN FIXATION IN LEGUMES

Nitrogenase can reduce protons as well as molecular nitrogen and the energy costs are the same, 4.5 ATP/2e. When H_2 is formed the process is wasteful if it escapes to the atmosphere. Some cells contain a hydrogenase capable of recycling H_2. Evans et al. (1977) reported great variation in the efficiency of nitrogen fixation. They also showed large differences in the efficiency of the soybean system which correlated with the strain of R. japonicum used to infect the plants. This group has now developed a method for measuring hydrogenase in R. japonicum in the free-living form and a spontaneous mutant which is incapable of using H_2. They are in good position to understand the interplay among the enzymes which control the efficiency of nitrogen fixation.

CHARACTERISTICS OF THE MOLYBDENUM IRON COFACTOR OF NITROGENASE

Nitrogenase is a molybdenum, iron containing metalloenyzme which catalyzes the synthesis of ammonia from molecular nitrogen.

Table 3. Physiological Parameters of Six F₃ Soybean Plants Produced from Seed of an F₂ Soybean Plant which Exhibited Green Leaves and Acetylene-reducing Nodules in the Field at a Time when Pods were Mature

Total dry wt. (g)	Pod dry wt. (g)	Pod dry matter (%)	Chlorophyll (mg/g, fresh wt.)	Leaf protein (mg/g, fresh wt.)	Ribulosebisphosphate carboxylase		Acetylene reduction (μmole C_2H_4 per hour per gram nodule dry wt.)
					Activity (μmole CO_2 per hour per gram fresh wt.)	Protein (mg/g, fresh wt.)	
74.4	30.8	32.8	1.3	11.7	36.1	2.1	1.4
67.4	28.0	32.6	1.6	16.1	54.4	3.3	0.7
63.5	25.6	31.6	1.4	14.0	45.8	1.9	1.0
84.1	27.5	32.1	0.5	6.1	20.6	0.9	0.7
60.0	20.9	28.7	0.5	5.1	16.9	0.7	0.1
52.4	19.8	29.1	0.4	4.9	12.0	0.5	0.1

S.S. Abu-Shakra et al. (1978) Science 199, 974. Copyright 1978 by the American Association for the Advancement of Science.

Nagatani et al. (1974) isolated a mutant of Azotobacter vinelandii
(UM 45) which contains inactive nitrogenase. In vitro the enzyme
is activated upon addition of an acid treated (inactive) nitrogenase. Then in 1977, Shah and Brill described the isolation of
a low molecular weight molybdenum containing factor from molybdenum-
iron protein which: (a) can activate UM 45 nitrogenase; (b) contains molybdenum, iron and labile sulfur; (c) is quite stable in
certain organic solvents. The iron-molybdenum cofactor from A.
vinelandii has been further purified and its physical and chemical
properties described (Newton et al., 1979). As isolated the
preparation is free of amino acids and has a ratio of 7 iron atoms
per molybdenum. Three redox states of the cofactor can be obtained
under appropriate conditions by titration with dithionite, methy-
lene blue and thiophenol. The reactions are reversible and can be
followed by characteristic EPR signals at 14°K. The cofactor also
reacts with EDTA with a loss of EPR signals. All four entities
are capable of acetylene reduction when added to UM 45 extracts.

NON-LEGUME, SYMBIOTIC NITROGEN-FIXING SYSTEMS: ALNUS

Among the non-legume nitrogen fixers are actinomycete-nodulated
woody perennial shrubs and trees. Alders are a part of this group.
They are fast growing trees and shrubs, they grow well in relatively
poor, nitrogen deficient soil and they appear to stimulate the
growth of other trees in silva-culture. For these reasons the
forestry industry is interested in them. In developing countries
there are additional needs - the need for renewable fuel, fiber
and biomass which make them candidates for further study.

Until recently these plants could only be experimentally
nodulated by using a crushed nodule inoculum. The endophyte could
not be grown ex planta in liquid or semi-solid media. In 1978
Callaham et al. reported the isolation and growth in culture of
the actinomycete (Frankia) which nodulates Comptonia peregrina.
Lalonde and Calvert (1979) have shown that this isolate can also
infect six species of Alnus producing nitrogen fixing nodules.
The endophyte can now be grown on a defined media and conditions
are reported for the inoculation of Alnus seeds and seedlings which
will produce large numbers of effective nitrogen-fixing nodules and
healthy seedlings. The potential exists for large scale production
of nodulated seedlings thus reducing the mortality at this critical
stage of growth.

NON-LEGUME, SYMBIOTIC NITROGEN-FIXING SYSTEMS: AZOLLA

One of the more unusual nitrogen-fixing symbiotic associations
is that which occurs between the water fern Azolla and its blue-
green alga symbiont, Anabaena azolla. The pair are found in fresh

water streams and ponds. Six species of the genus are known, they have world wide distribution. The combination is likely to have considerable agronomic impact in southeast Asia because it can be grown in dual culture with paddy rice and act as a slow release nitrogen fertilizer for it. In order to take maximum advantage of the potential which this association has, it is important to understand the structural and physiological characteristics which underlie its growth and development. Many of the important physiological characteristics have been summarized by Peters (1977).

Both partners are capable of photosynthesis. Their light absorbing characteristics are such that presumably more energy is captured by the association than by either individual partner. CO_2 compensation points are reproduced in Table 4. It can be seen that CO_2 fixation in the association is a more efficient process than in the fern itself or in (the distantly related) salvinia grown on nitrate. The alga exhibits the physiological characteristic of a lack of photorespiration. The absence of photorespiration is not due to the absence of ribulose bisphosphate oxygenase activity or incomplete photorespiration (Peters, 1977). Whatever the mechanism, the efficiency of the alga probably accounts for increased CO_2-fixing efficiency by the association.

The alga is the nitrogen-fixing partner and is capable of meeting the total N requirement of both partners. The fern can be grown in the absence of the alga, however fixed nitrogen must be added to the medium. Acetylene reduction and hydrogen production are shown in Figure 5. At a partial pressure of 0.1 atmospheres, acetylene is saturating for the acetylene reduction reaction and hydrogen production is almost completely inhibited. Nitrogenase activity is rapidly lost in most organisms in the presence of fixed nitrogen. In Azolla this does not appear to be the case, as shown in Table 5, where it can be seen that there is considerable acetylene reduction even after nine months (column c) compared with the activity after one month (column a).

Table 4. CO_2 Compensation Points (ppm CO_2) at 2% and 20% O_2

	2% O_2	20% O_2
Azolla with symbiont grown on N_2	19	51
Azolla with symbiont grown on $N_2 + NO_3^-$	20	55
Azolla without symbiont grown on NO_3^-	26	66
Isolated symbiont from N_2 grown Azolla	26	24
Salvinia grown on NO_3^-	38	75

Peters in "Genetic Engineering for Nitrogen Fixation" p. 246, 1977.

Table 5. Effect of Growth on Nitrate upon H_2 Production and C_2H_4 Reduction Values in parenthesis indicate separate experiment.

Assay Gas	(a) 400 ft-c		(b) 650 ft-c		(c) 400 ft-c		1200 ft-c
	H_2	C_2H_4	H_2	C_2H_4	H_2	C_2H_4	H_2
			nmoles/mg Chl·min				
$Ar-CO_2$	4.9	...	6.0 (8.2)	...	7.0	...	9.6
$Ar-C_2H_2-CO_2$	...	...	0.2 (0.2)	9.7 (12.6)	0.2	5.4	...
$Ar-C_2H_2-CO_2-CO$	...	...	6.0	0.2	...	...	...
$Ar-CO_2-O_2$	3.5	...	...	...	3.6	...	5.8
$Ar-C_2H_2-CO_2-O_2$	...	8.8	0.2	9.2	...	2.3	...
Air	...	...	1.4 (3.2)	...	...	...	...
Air-CO	...	...	2.7	...	...	...	...
$\dfrac{H_2(Ar-CO_2)*}{C_2H_4(Ar-C_2H_2-CO_2)}$	.55		.62 (.65)			1.30	

Incubations were for 16 to 18 hr at 26°C after 1 month (column a) and 9 months (column c) growth and for 5.5 to 6.5 hr at 28°C after 4 months (column b) growth, all on 4 mM NO_3^-. The assay medium contained 4 mM NO_3^-.

Peters in "Genetic Engineering for Nitrogen Fixation," p. 243, 1977.

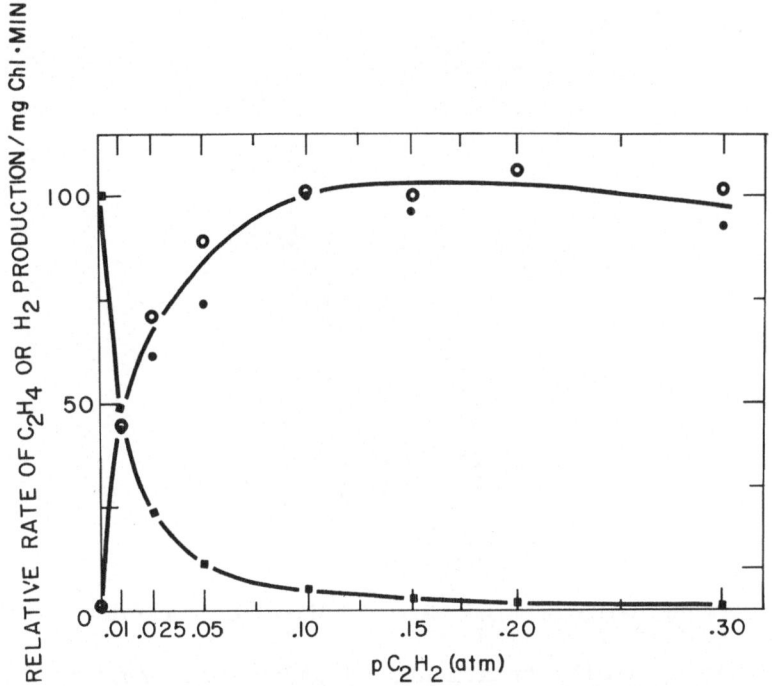

G.A.Peters, "Genetic Engineering for Nitrogen Fixation", p.240, 1977

Fig. 5. Effect of acetylene on nitrogenase catalyzed ethylene
production (0,●) and H_2 evolution (■).

     Azolla has been used as a green manure in Vietnam, Thailand,
and the Peoples Republic of China (Moore, 1969). In other areas of
southeast Asia it is not now part of the normal agricultural practice. Talley, Talley and Rains (1977) are studying the feasibility
of using Azolla in dual culture with rice to reduce the requirement
for N fertilizer in temperate-zone rice cultivation. They use two
indigenous species, A. filiculoides a frost tolerant species and
A. mexicana, a species which grows very well at elevated temperatures. Azolla is double cropped. During the fallow period for rice,
a crop of Azolla is grown and incorporated. Another crop is grown
during the rice growing season - the cover crop. Figure 6 shows
that it is possible to treble the yield of rice grain compared with
an unfertilized control by management of Azolla in this way. Table
6 shows that certain combinations maximize paddy yield. In other
data (personal communication), these workers have shown that as
much as 33% of N fertilizer can be replaced by an equivalent amount
of Azolla N with no loss in grain yield. Similar promising results
have been reported at the International Rice Research Institute
(Watanabe, 1978).

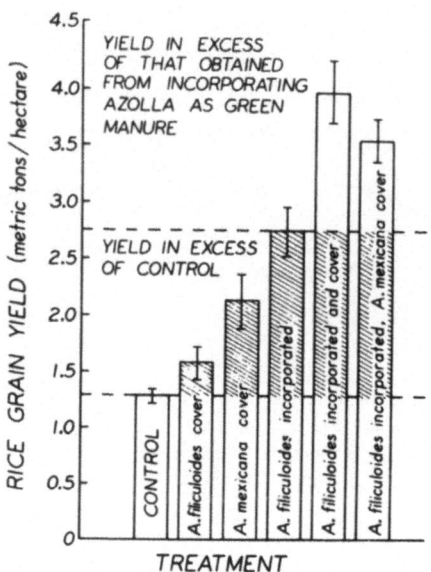

Talley, Talley and Rains "Genetic Engineering for Nitrogen
Fixation", p. 275, 1977

Fig. 6. Effect of <u>Azolla</u> treatments on paddy rice grain yield.
"<u>Azolla</u> cover" is grown in dual cultures with rice and
"<u>Azolla</u> Incorporated" is grown prior to rice planting and
is used as a green manure.

CONCLUDING REMARKS

While research activity and accomplishment in biological
nitrogen fixation has been good, most of the published work has
not been oriented to symbiotic associations of importance to developing countries. Among the leguminosae there is great diversity in
morphology, physiology and development. For example, both allantoin
and asparagine are circulating organic nitrogen compounds; soybeans
have allantoin while peas have the other. Because of this diversity
it is important to conduct research with seed grain legumes of
choice and the choice will differ from country-to-country.

Through 1975, biological nitrogen fixation research was an
activity that was not well supported by the federal government. In
fact, the only source of support for a non-university research
institution (such as the Kettering Laboratory) in this area of
research was through the Directorate for Biological Sciences of
the National Science Foundation. Increased funding became available
from the RANN division of NSF for problem-focused nitrogen fixation
research; in 1977 the Department of Agriculture launched its
competitive grants program which includes funds earmarked for

Table 6. Above-Soil Nitrogen Content of Rice and Weeds for Control and Azolla Test Paddies

Treatment	Nitrogen Content (kg/N/ha)							
	Rice Straw		Paddy Yield		Weeds		Total	
Control	11.1	0.9	13.9	1.0	Trace		24.9	0.1
A. filiculoides cover	13.0	1.3	16.9	1.4	Trace		29.8	2.6
A. mexicana cover	13.6	2.7	25.1	3.7	11.4	8.6	50.1	3.4
A. filiculoides Incorporated	15.9	0.6	26.8	2.4	13.7	3.3	56.5	1.5
A. filiculoides Incorporated A. filiculoides cover	21.2	2.1	42.0	2.6	3.4	1.4	66.6	4.5
A. filiculoides Incorporated A. mexicana cover	20.8	1.9	40.0	2.7	10.2	0.1	71.0	4.0

Talley, Talley and Rains in "Genetic Engineering for Nitrogen Fixation", p. 276, 1977.

nitrogen fixation research. Now that the ASRA division (formerly RANN) of NSF intends to phase out its support for nitrogen fixation research in this area we are increasingly dependent upon the U.S. Department of Agriculture competitive grants program for support of biological nitrogen fixation research. Clearly there is less than whole-hearted support for the competitive grants program by leaders of the agriculture research community. In this regard recent performance by Congress is cause for concern as well. Before the expiration of the first year granting period the House, on recommendation from the appropriations committee, voted no funds for the second year of the program. After House and Senate negotiation 15 million was appropriated for the second year. While there continues to be strong support from NSF's biological sciences directorate, their financial resources and the scope of their science support program militate against a large sustained commitment to this area. Financial restrictions have in the past limited the scope and the productivity of this program. Without the assurance of sustained, continuing long term support we will be unable to attract the available talented scientists needed to bring this program to maturity. Without funding there is minimal chance for making a significant contribution to increasing food production.

With sustained support, new knowledge from the basic sciences is
likely which can be translated through application to increased
production.

There is much to be gained from the multidisciplinary so-
called mission oriented approach in attacking a research problem
of this complexity. Insight gained from one approach facilitates
success in another. For maximum benefit, the agricultural sciences
should be part of the team. Students who are training for a
career in the agriculture sciences should, as part of their day-
to-day training, maintain close contact with chemists, biologists
and physicists. Basic sciences research activities should be in
the school of agriculture - preferably in the departments of
agronomy, horticulture, etc. working on agricultural problems.
Only in this way can we expect to break out of the stereotype of
"them-and-us". The problem is a complex one and all approaches
to increasing food production should be encouraged.

It is far too soon to be pessimistic about the future. One
needs only to compare food grain yields of a developed country
with the yield of a developing country to realize the vast potential for production which exists. Developing countries have the
added advantage of hindsight wisdom, learning from both the successes and the mistakes of the developed nations. This is not the
time for hand wringing, this is the time for problem solving.

REFERENCES

Abu-Shakra, S.S., Phillips, D.A., Huffaker, R.C., 1978, Nitrogen
    fixation and delayed leaf senescence in soybeans, _Science_,
    199:973-974.
Andersen, K., Shanmugam, K.T., and Valentine, R.C., 1977, Nitrogen
    fixation (NIF) regulatory mutants of _Klebsiella_: Determination
    of the energy cost of N_2 fixation _in vivo_, _in_: "Genetic
    Engineering for Nitrogen Fixation," A. Hollaender, ed., Plenum
    Press, New York, p. 95-107.
Bauer, W.D., Bhuvaneswari, T.V., Mort, A.J., and Turgeon, G., 1979,
    An initiation of infections in soybean by _Rhizobium_, _Plant
    Physiol_, in press.
Bergersen, F.J., and Turner, G.L., 1967, Nitrogen fixation by the
    bacteroid fraction of breis of soybean root nodules, _Biochem.
    Biophys. Acta_, 141:507.
Bergersen, F.J., and Turner, G.L., 1978, Activity of nitrogenase
    and glutamine synthetase in relation to availability of
    oxygen in continuous cultures of a strain of cowpea _Rhizobium_
    sp. supplied with excess ammonium, _Biochem. Biophys. Acta_,
    538:406.

Beringer, J.E., Beynon, J.L., Buchanan-Wollaston, A.V., and Johnston, A.W.B., 1978, Transfer of the drug-resistance transposon Tn 5 to Rhizobium, Nature, 276:633.

Beringer, J.E., Hoggan, S.A., and Johnston, A.W.B, 1978, Linkage mapping in R. leguminosarum by means of R plasmid mediated recombination, J. Gen. Microbiol., 104:201.

Bethlenfalvay, G.J., Abu-Shakra, S.S., and Phillips, D.A., 1978a, Interdependence of nitrogen nutrition and photosynthesis in Pisum sativum L. I. Effect of combined nitrogen on symbiotic nitrogen fixation and photosynthesis, Plant Physiol., 62:127.

Bethlenfalvay, G.J., Abu-Shakra, S.S., and Phillips, D.A., 1978b, Interdependence of nitrogen nutrition and photosynthesis in Pisus sativum L. II. Host plant response to nitrogen fixation by Rhizobium strains, Plant Physiol., 62:131.

Bhuvaneswari, T.V., Pueppke, S.G., and Bauer, W.D., 1977, Role of lectins in plant-microorganism interactions. I. Binding of soybean lectin to rhizobia, Plant Physiol., 60:486.

Bohlool, B.B., and Schmidt, E.L., 1974, Lectins: a possible basis for specificity in Rhizobium-legume root nodule symbiosis, Science, 185:269.

Callaham, D., Del Tredici, P., and Torrey, J.G., 1978, Isolation and cultivation in vitro of the actinomycete causing root nodulation in Comptonia, Science, 199:899.

Ching, T.M., Hedtke, S., and Newcomb, W., 1977, Isolation of bacteria, transforming bacteria and bacteroids from soybean nodules, Plant Physiol., 60:771.

Darrow, R.A., and Knotts, R.R., 1977, Two forms of glutamine synthetase in free-living root-nodule bacteria, Biochem. Biophys. Res. Commun., 78:554.

Dazzo, F.B., and Brill, W.J., 1977, Receptor site on clover and alfalfa for Rhizobium, Applied Environ. Microbiology, 33:132.

Dazzo, F.B., and Hubbell, D.H., 1975, Cross-reactive antigens and lectin on determinants of symbiotic specificity in Rhizobium-clover association, Applied Microbiology, 30:1017.

Evans, H.J., Ruiz-Argüeso, T., Jennings, N., and Hanus, J., 1977, Energy coupling efficiency of symbiotic nitrogen fixation, in: "Genetic Engineering for Nitrogen Fixation," A. Hollaender, ed., Plenum Press, New York, p. 333.

Fujihara, S., Yamamoto, K. and Yamaguchi, M., 1977, A possible role of allantoin and the influence of nodulation on its production in soybean plants, Plant Soil, 48:233-242.

Gardner, I.C., and Leaf, G., 1960, Translocation of citrulline in Alnus glutinosa, Plant Physiol., 35:948-950.

Ginsburg, A., and Stadtman, E.R., 1973, Regulation of glutamine synthesis in Escherichia coli, in: "The Enzymes of Glutamine Metabolism," S.M. Prusiner and E.R. Stadtman, eds., Academic Press, New York, pp. 9-43.

Hardy, R.W.F., and Havelka, U.D., 1975, Nitrogen fixation research: A key to world food? Science, 188:633.

Hardy, R.W.F, and Havelka, U.D., 1975, Photosynthate as a major
    factor limiting nitrogen fixation by field-grown legumes with
    emphasis on soybeans, in: "Symbiotic Nitrogen Fixation in
    Plants," P. Nutman, ed., International Biological Programme
    Series, Vol. 7, Cambridge University Press, London.
Herridge, D.F., Atkins, C.A., Pate, J.A., and Rainbird, R.M., 1978,
    Allantoin and allantoic acid in the nitrogen economy of the
    cowpea, Plant Physiol., 62:495.
Herridge, D.E., and Pate J.S., 1977, Utilization of net photosynthate
    for nitrogen fixation and protein production in an annual
    legume, Plant Physiol., 60:759-764.
Hooykaas, P.J.J., Klapwijk, P.M., Nuti, M.P., Schilperoort, R.A.,
    and Rörsch, A., 1977, Transfer of the Agrobacterium tumefaciens
    TI plasmid to avirulent agrobacteria and to Rhizobium ex planta,
    J. Gen. Microbiol., 98:477.
Johnston, A.W.B., Benyon, J.L., Buchanan-Wollaston, A.V., Setchell,
    S.M., Hirsch, P.R., and Beringer, J.E., 1978, High frequency
    transfer of nodulating ability between strains and species of
    Rhizobium, Nature, 276:635.
Keister, D.L., 1975, Acetylene reduction by pure cultures of
    rhizobia, J. Bacteriol., 123:1265.
Klucas, R.V., and Burris, R.H., 1967, Locus of nitrogen fixation in
    soybean nodules. Fixation by crushed nodules, Biochem. Biophys.
    Acta, 136:399.
Kondorosi, A., Kiss, G.B., Forrai, T., Vincze, E., and Banfalvi, Z.,
    1977, Circular linkage map of R. meliloti chromosome, Nature,
    268:525.
Kurz, W.G.W., and LaRue, T.A., 1975, Nitrogenase activity in
    rhizobia in absence of plant host, Nature, 256:407.
Lalonde, M., and Calvert, H.E., 1979. Production of Frankia hyphae
    and spores as an infective inoculant for Alnus species, in:
    "Proceedings of Symbiotic Nitrogen Fixation in the Management
    of Temperate Forests," Corvallis, OR.
Lewis, O.A.M., and Pate, J.S., 1973, The significance of transpira-
    tionally derived nitrogen in protein synthesis in fruiting
    plants of pea, J. Expt. Bot., 24:596.
Magasanik, B., Prival, M.J., Brenchley, J.E., Tyler, B.M., DeLeo,
    A.B., Streicher, S.L., Bender, R.A., and Paris, C.G., 1974,
    Glutamine synthetase as a regulator of enzyme synthesis, in:
    "Current Topics in Cellular Regulation," B.L. Horecker and
    E.R. Stadtman, eds., Academic Press, New York, Vol. 8,
    pp. 119-138.
Maier, R.J., Campbell, N.E.R., Hanus, F.J., Simpson, F.B., Russell,
    S.A., and Evans, H.J., 1979, Expression of hydrogenase
    activity in free-living Rhizobium japonicum, Proc. Nat.
    Acad. Sci., US, 75:3258.
Matsumoto, T., Yatazawa, M., and Yamamoto, Y., 1977, Distribution
    and change in the contents of allantoin and allantoic acid in
    developing nodulating and non-nodulating soybean plants,
    Plant Cell Physiol., 18:353.

Matsumoto, T., Yatazawa, M., and Yamamoto, Y., 1977, Incorporation of ^{15}N into allantoin in nodulated soybean plants supplied with ^{15}N, Plant Cell Physiol., 18:459.

Matsumoto, T., Yatazawa, M., and Yamamoto, Y., 1977, Effects of exogenous nitrogen compounds on the concentrations of allantoin and various constituents in several organs of soybean plants, Plant Cell Physiol., 18:613.

McCombs, J.A., Elliott, J., and Dilworth, M.J., 1975, Acetylene reduction by Rhizobium in pure culture, Nature, 256:409.

Meade, H.M., and Signer, E.R., 1977, Genetic mapping of Rhizobium meliloti, Proc. Nat. Acad. Sci., US, 74:2076.

Meadows, D.H., Meadows, D.L., Randers, J., and Behreus III, W.W., 1972, The limits to growth, Signet Books, New York, New York.

Moore, A.W., 1969, Azolla: biology and agronomic significance, Bot. Rev., 35:17.

Mothes, K., 1961, The metabolism of urea and ureides, Can. J. Bot., 39:1785.

Nagatani, H.H., Shah, V.K., and Brill, W.J., 1974, Activation of inactive nitrogenase by acid-treated component. I. J. Bacteriol., 120:697.

Newton, W.E., Burgess, B.K., and Stiefel, E.I., 1979, Chemical Properties of the Fe-Mo cofactor from nitrogenase, in: "Molybdenum Chemistry of Biological Significance," Plenum Press.

Nuti, M.P., Ledeboer, A.M., Lepidi, A.A., and Schilperoort, R.A., 1977, Large plasmids in different Rhizobium species, J. Gen. Microbiol., 100:241-248.

O'Gara, F., and Shanmugam, K.T., 1976, Regulation of nitrogen fixation by rhizobia export of fixed nitrogen as ammonium ion, Biochem. Biophys. Acta, 437, 437:313.

Pagen, J.D., Child, J.J., Scowcroft, W.R., and Gibson, A.H., 1975, Nitrogen fixation by Rhizobium cultured on a defined medium, Nature, 256:407.

Pate, J.S., Sharkey, P.J., and Atkins, C.A., 1977, Nutrition of a developing legume fruit. Functional economy in terms of carbon, nitrogen and water, Plant Physiol., 59:506.

Peters, G.A., 1977, The Azolla-Anabaena azolla symbiosis, in: "Genetic Engineering for Nitrogen Fixation," A. Hollaender, ed., Plenum Press, New York, pp. 231-258.

Rao, V.R., Darrow, R.A., and Keister, D.L., 1978, Effect of oxygen tension on nitrogenase and on glutamine synthetases I and II in Rhizobium japonicum 61A76, Biochem. Biophys. Res. Commun., 81:224-231.

Ray, T.B., Mayne, B.C., Toia, R.E. Jr., and Peters, G.A., 1979, personal communication.

Reporter, M., and Bednarski, M.A., 1978, Expression of rhizobial nitrogenase: The influence of plant cell conditioned medium, Applied Environ. Microbiol., 36:115.

Shah, V.K., and Brill, W.J., 1977, Isolation of an iron-molybdenum cofactor from nitrogenase, Proc. Nat. Acad. Sci., US, 74:3249.

Shanmugam, K.T., O'Gara, F. Andersen, K., and Velentine, R.C., 1978, Biological nitrogen fixation, Am. Rev. Plant Physiol., 29:263.

Skotnicki, M., Rolfe, B., and Reporter, M., 1978, Plant peptido-glucans affect nitrogenase activity and oxidative phosphorylation in Rhizobium trifolii strain T1, Biophys. J., 25:275a.

Skotnicki, M.L., Rolfe, B.G., and Reporter, M., 1979, Nitrogenase activity in pure cultures of spectinomycin-resistant fast and slow-growing Rhizobium, Biochem. Biophys. Res. Commun., 86:975.

Storey, R., and Reporter, M., 1978, H_2 production and C_2H_2 reduction in rhizobia influenced by plant substances, Plant. Physiol., 61S:8.

Streeter, J.G., 1979, Allantoin and allantoic acid in tissues and stem exudate from field-grown soybean plants, Plant Physiol., 63:478.

Talley, S.N., Talley, B.S., and Rains, D.W., 1977, Nitrogen fixation by Azolla in rice in fields, in: "Genetic Engineering for Nitrogen Fixation," A. Hollaender, ed., Plenum Press, N.Y., p. 259.

Tjepkema, J.D., and Evans, J.J., 1975, Nitrogen fixation by free-living Rhizobium in a defined liquid medium, Biochem. Biophys. Res. Commun., 65:625.

Tubbs, R.S., 1976, Regulation of nitrogen fixation in Rhizobium-sp., Applied Environ. Microbiol., 32:483-488.

Upchurch, R.G., and Elkan, G.H., 1978, Ammonia assimilation in Rhizobium japonicum colonial derivatives differing in nitrogen-fixing efficiency, J. Gen. Microbiol., 104:219.

Valentine, R.C., 1977, Genetic engineering of nitrogen fixation, in: "Genetic Engineering for Nitrogen Fixation," A. Hollaender, ed., Plenum Press, New York, NY, p. 496.

Watanabe, I., 1978, Azolla and its use in lowland rice culture, Soil and Microbes (Japan), 20:1-10.

Watanabe, I., Espinas, C.R., Berja, N.S., and Alimagno, B.V., 1977, Utilization of the Azolla-Anabaena azolla complex as a nitrogen fertilizer for rice, IRRI Research Paper Series, No. 11, The International Rice Research Institute, Manila, Philippines.

Watt, G.D., Bulen, W.A., Burns, A., and Hadfield, K.L., 1975, Stoichiometry ATP/2e values and energy requirements for reactions catalyzed by nitrogenase from Azotobacter vinelandii, Biochem., 14:4266-4272.

Wohlhueter, R.M., Schutt, H., and Holzer, H., 1973, Regulation of glutamine synthesis in vivo in E. coli, in: "The Enzymes of Glutamine Metabolism," S. Prusiner and E.R. Stadtman, eds., Academic Press, New York, pp. 45-64.

TRANSLATING BASIC RESEARCH ON BIOLOGICAL NITROGEN FIXATION TO

IMPROVED CROP PRODUCTION IN LESS-DEVELOPED COUNTRIES -- A USER'S

VIEW

R.W.F. Hardy

Central Research & Development Department
Experimental Station
E.I. du Pont de Nemours & Company
Wilmington, DE 19898

INTRODUCTION

The importance of nitrogen in crop production was animated almost two decades ago in a New Yorker cartoon and earlier this year in a Farm Chemicals cartoon (Anon., 1979a). In the simple but dramatic New Yorker cartoon, a single human-like plant in a desert environment was gasping "Ammonia, Ammonia". In the more complex Farm Chemicals cartoon sequence, the assumed human-like plant was treated in the style of the 1970's by playing its favorite records and participation in group sessions with nature lovers while the fundamental nutritional need "some nitrogen would help" was ignored. The identical message of both cartoons is as relevant in 1979 as it was in 1960 and is of at least as much relevance to less-developed as to more-developed countries. Statistical documentation of the message of the cartoons is the direct relationship between the increasing yield of cereal grains and the increasing use of fertilizer nitrogen from 1950 to 1975 for the less-developed and the more-developed countries (Hardy and Havelka, 1975). In this article I will consider the problem of future nitrogen need for crop production and approaches to meeting this need. Emphasis will be placed on less-developed countries to the extent practical, but many of the solutions are common for both more- and less-developed countries.

GENERAL APPROACHES TO MEETING NITROGEN NEEDS

The world need for cereal grain production has been estimated to double from 1300 MM tons in 1975 to 2600 MM tons in 2000 A.D. (Hardy, 1977). The need for nitrogen in cereal grain will increase from 39 to 78 MM tons based on the above increase in production and the following assumptions:

1) The average cereal grain contains 2% N, and

2) Two-thirds of total plant N is in the grain at harvest.

General approaches to meet this need include:

1) An increase of world N fertilizer production from the current approximately 50 MM tons to about 160 MM tons by 2000 A.D.,

2) improve the efficiency of fertilizer nitrogen use by cereal grains from a current average of about 50% to 75%,

3) develop a practical process for abiological N_2 fixation with zero direct energy input such as an oxidative rather than the current Haber-Bosch process which is reductive, and/or

4) develop a biological N_2 fixation system for cereal grains that is compatible with high yield.

World ammonia capacity was rated at 85 MM tons in 1977-78, and it has been projected to increase at a rate of 6.4% per year to 116 MM tons by 1982-83 (Table 1) (Anon., 1979a). Continuation of

Table 1.   World Ammonia Capacity by Regions

	1977/78	1982/83	Capacity	Consumption
	Million Tons/Yr		%/Yr	%/Yr
North America	19	20	1	4
Western Europe/Japan	20	22	2	3
Eastern Europe/USSR	23	36	9	7
Developing Countries/ China	23	38	10	8
World	85	116	6.4	6.0

Adopted from Anon., 1979a.

this trend should provide adequate capacity to meet the estimated 160 MM ton need by 2000 A.D. Moreover, this capacity is increasing most rapidly at a rate of 10% per year in the developing countries including China where there is an increase in consumption of 8% per year, and where the need is the greatest.

Rice is a major cereal grain crop in less-developed countries and is also a major consumer of fertilizer nitrogen in less-developed countries such as those in Southeast Asia (Table 2) (Anon., 1979b). Efficiency of fertilizer nitrogen use is an unacceptable 30% for rice so that improvement in the efficiency of use in this crop would produce a major effect on nitrogen needs. Nitrification inhibitors such as N-Serve may be effective in increasing the efficiency of use of ammonia fertilizers (Huber et al., 1977).

A new process for abiological N_2 fixation with zero direct energy input would be the most significant advance that I can envision for the provision of world crops with adequate nitrogen. Unfortunately, the scientific base for such a process is almost nonexistent since most of the exploratory chemistry on N_2 fixation during the last 20 years has focused on reductive processes. An exception is work at the Kettering Research Laboratory on an oxidative albeit high energy process in which the energy is supplied by a renewable resource such as wind. Additional limited tests are being conducted in India and elsewhere in the U.S. It should be possible to discover oxidative processes with much less or possibly even no direct requirement for energy, and innovative research is needed in this area. The current annual value of world nitrogen fertilizer of about $10 billion indicates the economic opportunities for a new process and, of course, this will grow as the use of fertilizer nitrogen grows.

Several biological N_2 fixation systems for cereal grains can be suggested such as the so-called associative symbioses, e.g., Azospirillum-cereal grain or grass, nodulated cereal grains, and cereal grains into whose chromosomes have been incorporated the genetic information for N_2 fixation. The initial high expectations of some of these such as the associative symbioses have not been substantiated in more critical tests (Evans and Barber, 1977), and there are many advances that must be made before genetic engineering of higher plants will be a reality (Shanmugan and Valentine, 1975). In addition, the biological system must be compatible with high yield, and unmodified biological N_2 fixation may be too excessively energy consuming to meet this requirement. The Azolla-Anabaena symbiosis which is used in many paddy rice fields in the Far East is probably the single case where a biological N_2 fixation system already exists that is of substantial importance for a cereal grain, and Dr. Lamborg's paper discusses basic research on this system.

Table 2.   Rice is Major Fertilizer User in Developing Countries
            of Southeast Asia

	Consumption		Production
	Rice	Total	
	(Thousand Tons N)		
India	516	2149	1508
South Korea	253	468	212
Indonesia	215	342	207
Vietnam	194	194	0
Bangladesh	147	147	131
Taiwan	139	231	212
Philippines	101	144	62
West Malaysia	68	72	34
Thailand	45	79	4
Pakistan	40	444	316
Sri Lanka	25	38	0
Burma	21	42	47
Nepal	8	8	0
Afghanistan	1	28	50
Laos	NA	NA	NA
Khmer Republic	NA	NA	NA
Total	1773	4836	3112

Adopted from Anon., 1979b.

The world need for grain legumes is estimated to quadruple
from 130 MM tons in 1975 to 520 MM tons in 2000 A.D. (Hardy, 1977;
Hardy et al., 1979). The need for N in grain legumes will increase
from 10 to 39 MM tons based on the above increase in production and
the following assumptions:

1)   grain legume contains 5% N in the grain, and

2)   the grain contains two-thirds of the total plant N.

General approaches to meet this need include:

1)   enhanced symbiotic biological N_2 fixation, and/or

2)   development of fertilizer responsive systems for legumes.

World grain legumes are a conglomerate dominated by soybeans
which represent about 50% of total production (Table 3) (Anon.,
1976). Major driving forces to support a quadrupling of the need
for grain legumes by 2000 A.D. are the growing demand for feed
protein and for several legumes consumed directly as food. Meat

Table 3.   World Grain Legumes -- Production, Area, and Yield

	Production (10 Metric Tons)	Area (10^6 ha)	Yield (kg/ha)
Soybeans	68.4	46.5	1471
Groundnuts	19.1	19.4	986
Dry Beans	13.3	24.7	535
Dry Peas	10.6	10.6	999
Chick Peas	5.7	9.7	590
Broad Beans	6.4	5.4	1166
Pigeon Peas	2.0	2.7	699
Vetches	1.6	1.6	997
Lentils	1.2	1.9	640
Cow Peas	1.1	5.2	212
Lupins	0.6	0.9	647
Other Pulses	3.7	6.8	543
Total or Average	133.7	135.4	990

Adopted from Anon., 1976.

consumption in more-developed countries is increasing at 4% per year while that in less-developed countries is increasing at 3.7% (Table 4) (Boerma, 1979). Total world protein meal consumption increased by 44 MM tons from 40 to 84 MM tons from 1965-79 while soybean meal increased by 40 MM tons during the same period which accounts for almost all of the increase (Hardy et al., 1979). Such statistics strongly suggest that there is a high growth rate in the demand for grain legumes especially soybeans.

The most realistic opportunity for increasing biological N_2 fixation is to enhance N_2 fixation by the Rhizobium-legume symbiosis since only improvement of an existing system is needed rather than creation of a new system such as noted above for the cereal grains. Several approaches which are feasible include an improved inoculation technology, improved strains of Rhizobium, improved plant cultivars, and a better matching of the Rhizobium strain to the legume cultivar. Energy provision or minimization as discussed below is a major opportunity. The development of nitrogen fertilizer systems to which legumes respond is an alternate or a supplement to enhanced biological N_2 fixation. Most legumes show little benefit from the application of other than starter fertilizer nitrogen -- soybeans in the U.S. receive an average of only 15 kg fertilizer N/ha while corn receives about 130 kg N/ha (Hardy and Havelka, 1978). However, results in recent years show that foliar fertilization of soybeans during pod filling with mixtures of N-P-K-S similar to their occurrence in mature seeds gives substantial

Table 4.   Meat Production in Developed and Developing Countries

| | 1948/52 | 1976 | Growth |
	Million Metric Tons		%/Yr
Developed Countries	32.3	81.9	4.0
Developing Countries	16.5	39.7	3.7
World	48.8	121.6	3.9

Adopted from Boerma, 1979.

yield increases in some tests (Garcia and Hanway, 1976), although
reproducibility was less than satisfactory. Further work to better
define this exciting approach may provide a practical route to
increase nitrogen input and yield of grain legumes.

ADVANCES AND IMPACTS OF NITROGEN INPUT RESEARCH

     Research in almost all disciplines related to nitrogen input
has been highly productive of basic information during the past
20 years as summarized in Table 5. Examples include the extraction,
purification and characterization of nitrogenase, the C_2H_2-C_2H_4
assay, formation of transition metal·N_2 complexes and their
reduction under "mild" conditions, molecular genetics of the nif
genes and their cloning, N_2 fixation by cultured Rhizobium, the
ATP requirement and high energy cost of biological N_2 fixation,
nitrification inhibitors, high-lysine corn and sorghum, and photosynthate limitation for N_2 fixation by legumes. Only a few of
these advances have impacted nitrogen input practically and almost
all of these practical impacts do not involve biological N_2 fixation.
Examples include 1000-ton per day nitrogen fixation plants made
possible by engineering advances, nitrification inhibitors made
possible by chemical synthesis coupled with empirical screening
and high-lysine corn and sorghum made possible by plant breeding.
Unfortunately, the dynamic basic advances in both abiological and
biological N_2 fixation have not been accompanied by a practical
development with the possible exception of the C_2H_2-C_2H_4 assay.
Hopefully the record will improve during the next 20 years since
the basic advances as discussed below have revealed many weaknesses
and consequently opportunities for improvement of the biological
N_2 fixation system.

THE WHAT'S WRONG WITH OR OPPORTUNITIES FOR IMPROVEMENT OF BIOLOGICAL
N_2 FIXATION

     During recent years, I have tabulated a list (Hardy and
Havelka, 1975; Hardy et al., 1979; Hardy, 1979) of the limitations

Table 5.   <u>Research Advances in Nitrogen Input Research</u>

(Only Those in Dotted Enclosures Have Practical Utility)

Year	MATHEMATICS & ENGINEERING	CHEMISTRY	BIOCHEMISTRY	GENETICS	BIOLOGY	AGRONOMY
1980	Rhizobium-Cereal Langmuir Adsorption Isotherm			H$_2$ Uptake Plasmid	Ureides & Asparagine Assimilates	
				Mapping 14 nif Genes	CH$_2$O Cost for Legume N$_2$ Fixation	⌐Azolla-Anabaena Rediscovery⌐ (dotted)
		N$_2$ $\xrightarrow{\text{h}\nu}{\text{TiO}_2}$ NH$_3$				
					H$_2$ Uptake Hydrogenase	
			N$_2$H$_2$ Intermediate	Cloning nif Genes	Plant Lectin Rhizobia Interaction	"A" Value Method
1975		M(N$_2$)$_2$ → N$_2$H$_4$ or NH$_3$	Fe-Mo Cofactor		N$_2$ Fixation by Cultured Rhizobia	⌐Soybean Foliar Fertilization⌐ (dotted)
				nif Regulation Glutamine Synthetase		Photosynthate Limitation for Legume N$_2$ Fixation
			Fe Protein-ATP-e Mo-Fe Protein Interaction	nif Transfer		
1970		Molybdo-Thiol Reductant	Mo-Fe Active Site			
			Crystallization of Mo-Fe Protein			
			⌐C$_2$H$_2$-C$_2$H$_4$ Assay⌐ (dotted)			
1965		M(N$_2$) Complexes	Substrate Versatility			Associative Symbioses
			ATP-Requirement of Nitrogenase			
	⌐1000+ Ton/Day N$_2$ Fixation Plants⌐ (dotted)	RMgBr + (C$_5$H$_5$)$_2$ TiCl$_2$				
						⌐High-Lysine Corn⌐ (dotted)
1960		⌐Nitrification Inhibitor⌐ (dotted)	Nitrogenase Extracted			

or "what's wrongs" with the biological N_2 fixation system. These
"what's wrongs" identify opportunities to improve the biological
N_2 fixation system through chemical, genetic, physical, cultural,
or policy approaches. This list expands as our basic knowledge
of the system expands, and recent highlights are presented in
Dr. Lamborg's paper. The current list (Table 6) which includes
the areas of mathematics, chemistry, biochemistry, genetics,
biology, agronomy, and policy contains over 80 "what's wrongs."
Several limitations are of scientific interest only while others
are judged to represent opportunities for significant improvement.
In these latter cases, the chances for success range from extremely
low, e.g., reduction of the high direct ATP requirement of nitrogenase or elimination of H^+ reduction by nitrogenase, to more
probable, e.g., substantial improvement in the efficiency of
carbohydrate use, to highly probable, e.g., improved inoculation
technology. Specific steps in legume N_2 fixation will be outlined
to identify the state of knowledge, the biological cost of the
process, and the potential opportunities for improvement.

STEPS IN LEGUME N_2 FIXATION

The legume N_2 fixation system will be considered as the product
of six steps. The initial step is production of photosynthate in
the leaf. Knowledge of this system has identified the competition
of O_2 with CO_2 and the resultant photorespiratory losses of CO_2
in so-called C_3 plants -- all legumes tested so far are C_3 plants
but it is reasonable to expect that C_4 legumes may be found in
tropical environments. CO_2 enrichment of the canopy of four field-
grown grain legumes from flowering to senescence substantially
increased N_2 fixation (Hardy and Havelka, 1977). Undoubtedly there
are opportunities other than decreasing photorespiration for the
improvement of photosynthate production and these are of at least
as great a significance to increasing N_2 fixation as the improvement
of the N_2 fixation system itself.

Sucrose, the major translocation product of photosynthesis must
be exported from the leaf and imported by the nodule to fuel the
N_2 fixation reaction. Basic information on the loading of sucrose
into the transport stream is being elucidated (Giaquinta, 1977).
Opportunities may be found to improve this process and to regulate
the sink strength of the nodule relative to that of the reproductive
and other vegetative parts of the plant.

The next step is the generation of ATP, reductant, and the carbon
skeletons needed for the fixation of N_2 in the nodule and export
of the fixed N_2 from the nodule. Recent experiments in our
laboratory have measured the overall cost of N_2 fixation by comparison of root or nodulated root respiration of nodulating and non-

Table 6. What's Wrong With Biological N_2 Fixation

Technical

Mathematical (1)
        No treatment for Rhizobium-Legume or Azospirillum-Cereal

Chemical (1)
        *No Chemical Effectors

Biochemical (34)
        *Minimal Variation in Nitrogenases
         Large Molecular Weight
         Multicomponent
         Activating Factor in Some Cases
         Complex Fe-Mo Cofactor
         Processing of Precursor Proteins and Insertion of Fe-Mo
          Cofactor
         Optimum Component Ratio
         Allosteric Characterization
        *O_2 Lability of Each Component
         O_2 Inhibition of Reaction
         Special O_2 Handling Molecules and Systems, e.g., LHb
         Temperature Instability
         Biphasic Arahenius Plot with High Apparent Activation
          Energy at $\leq$ 18-20°
         Mo, Fe, and S Content
         Systems for Uptake and Storage of Mo and Fe
        *Low Turnover (50-100)
        *Electrons for Reduction
         Special Electron Donors - Fd and Fld
         High Redox Potential (NADPH/NADP > 100)
        *High Direct ATP Requirement (4 ATP's/2e)
         ADP Inhibition
         High ATP:ADP Ration (> 10)
         High ATP-Generating Capacity
        *Substrate Promiscuity - H^+ Reduction
         Uptake Hydrogenase Needed
         Regulating Molecules for Nitrogenase and Process
         Inhibition by NO_3- or NO_3- Product
         Ammonia Exporting System in Microsymbiont
         Special NH_3 Incorporating Systems and Products - Ureides
         Multiple Forms of Glutamine Synthetase
         Ancillary Molecules for Formation of Symbiosis-Lectins
          (trifoliin), Cell Wall Polysaccharides
         Accumulation of β-Hydroxybutyrate Polymer
         CH_2O Substrate Used by Bacteroid?
         CO_2 Refixation by PEP Carboxylase - Significance?

Genetic (19)
        *No Useful Variation in Genes for Structural Proteins
        *Large Number of Genes for nif (14)

Table 6 (cont.)

    Genetic (cont.)
        Large Size of Above Genes (30 kbases)
        Number of Copies of <u>nif</u> Genes?
        Complex Regulation
      *Prokaryotic Limitation
       Plasmid or Chromosonal Location of <u>nif</u> in <u>Rhizobium</u>
       Other Genes in Addition to <u>nif</u> to Form Symbioses
       Both Macrosymbiont and Microsymbiont Genes
       Communication Between Bacteriod and Host?
       Higher DNA Content in Bacteroid than Bacteria
       Multiple Copies of cDNA for LHb
       mRNA for LHb Larger than Required
      *Juvenility of <u>Rhizobium</u> Molecular Genetics for <u>nif</u>
       Minimal Information on Identification and Location of
        Genes for Symbioses
      *Similarity of <u>Agrobacterium</u> and <u>Rhizobium</u> (Symbiotic
       vs. Pathogenic)
       Regulation of Nitrogenase Expression
       Optimum Matching of Macrosymbiont and Microsymbiont Genes
       Claims of Super Mutants?

    <u>Biological and Agronomic</u> (26)
      *Limited Occurence
       Low pO_2 Required in Many Cases
       Low Activity in Isolated Bacteroid
      *Asymbiotic Fixation by Only Some <u>Rhizobium</u> and Fixation
       is Usually Much Less than in Symbiosis
       Membrane to Separate Host and <u>Rhizobium</u>
      *Fixed N (NO_3-, NH_3) Inhibits Infection, Nodulation,
       Fixation and Induces Senescence
       Denitrification in <u>Rhizobium</u> Bacteroid?
       Role of Capsular, Lipo- and Exopolysaccharide in
        Specification
       Essentiality of Lectins (Plant) for Infection
       Host-<u>Rhizobium</u> Specificity
       Cellulase (Host) and Pectinase (Bacteria) for Cell Wall
        Dissolution
      *High and Inefficient Use of Carbohydrate (10 kg CH_2O/kg
       N_2 Fixed)
      *Evolution of H_2 by Several <u>Rhizobium</u>
      *Photosynthetic Inefficiency of Legumes and Most Cereals
      *Time Profile of N_2 Fixation Does Not Match Need of Soybeans
      *Premature Senescence
      *Inadequate Amount for High Yield Soybeans
      *Inhibition by Fixed N as Occurs in High Fertility Soils
      *High Cost for Synthesis and Maintenance of Nodules
      *Need for a Multiplicity of Strains Dependent on Cultivars,
       Soil, Climate

Table 6 (Cont.)

<u>Biological and Agronomic</u> (26)
    *Problem of Manufacture, Storage, Handling, and
     Application of Labile <u>Rhizobium</u>
    *Competition Between Applied and Endogenous <u>Rhizobium</u>
     (Emphasis on Greater Competitiveness May Not be
     Desirable)
    Inefficient Coupling to Plant with Free-Living Diazotrophs
     or Associations Not Involving the Plant
    *Insignificant for Most Cereals (1-10 kg N/ha·yr) with
     Possible Exception of Rice and <u>Azolla-Anabaena</u>
    Susceptibility to Pathogens
    Measurement Techniques - Indirect vs. Direct - Kinetic
     vs. Integrative

<u>Policy</u>

    <u>Proprietariness</u> (1)
       *Inadequate and Inherently Difficult to Encourage
        Adequate Investment in Exploration, Development, and
        Implementation of Significant Solutions

    <u>Quality</u> (1)
       *Quality Control of Product and Use

---

*Judged to be most significant limitation.
Adopted from Hardy, 1979.

nodulating isolines of soybeans (Heytler and Hardy, 1979). The
calculated costs range from a high of 100 grams to a low of 10 or
less grams of carbohydrate oxidized to CO_2 per gram of N_2 fixed
to NH_3. The high value is associated with very low rates of N_2
fixation per plant while the low value is associated with high
rates per plant. Thus a high rate of N_2 fixation per plant is
most efficient and the divergent costs of N_2 fixation reported in
the literature may be attributable to different rates of N_2
fixation. About 280 grams of glucose are oxidized in the nodulated
root to produce ATP and reductant equivalent to 60 ATP's calculated
from the low value.

In the next step which is the nitrogenase reaction, 28 of the
above 60 ATP equivalents are used to reduce one mole of N_2 (28
grams) to two moles of NH_3 and concomitantly to produce one mole
of H_2 from protons in the aqueous medium. This calculation is
based on <u>in vitro</u> measurements of the nitrogenase reaction. Seventy-
five percent of the ATP equivalents consumed directly by nitrogenase are used to reduce N_2 and 25% are "wasted" to produce the
unwanted H_2. There is a major but undoubtedly difficult opportunity
to improve the efficiency of the N_2 fixation system by eliminating
H_2 production (Hardy and Havelka, 1977).

Recovery of some but less than 50% of the ATP equivalents lost in H_2 production can occur if the system contains an uptake hydrogenase so as to oxidize H_2 to H_2O and couple this oxidation with the generation of 2-3 ATP's per H_2 oxidized. Greenhouse experiments have recently shown that N_2 fixation in soybeans is significantly improved by inoculation with Rhizobium which contain an uptake hydrogenase compared to those that do not (Albrecht et al., 1979). Field data are needed. Clearly, an uptake hydrogenase system should be included in all Rhizobium used for inoculation.

Other reactions of the N_2 fixation system use one-half or more of the ATP equivalents. It is anticipated that additional opportunities to improve the efficiency of the system may be uncovered in this less well-understood but energy consuming part of the N_2-fixing system. The above information has been found mainly by studies on soybeans; however, it should be applicable to other legumes grown in the less-developed countries.

POSSIBLE FUTURE TECHNOLOGIES

A large number of future technologies for increasing nitrogen input may be suggested. They extend from the highly probable with impact anticipated in the near term to the highly speculative with impact only occurring in the long term.

Examples of highly probable short-term impact technologies are:

1) Multiple cropping of legumes and nonlegumes and practical examples already exist,

2) Effective rhizobial inoculation technology which is inadequate in more-developed countries and may be even more difficult for less-developed countries,

3) High efficiency of fixed nitrogen use by crops with rice a major opportunity in less-developed countries,

4) Nitrogen fertilizer responsive legumes or systems based on genetic as well as fertilizer approaches,

5) Increased harvest index for nitrogen which may have high relevance to some of the food legumes of less-developed countries where the harvested seed accounts for only a small portion of the plant,

6) Symbioses insensitive with respect to nodulation or to N_2 fixation to fixed nitrogen will be of most significance to soils with high nitrogen fertility and consequently will be of greater impact to areas of high use of fertilizer nitrogen,

7) Rhizobium with H_2 uptake gene should be applicable to all inoculum for both more- and less-developed countries, and

8) Improved Azolla-Anabaena symbiosis for paddy rice production.

Examples of less probable mid-term impact technologies are:

1) Promiscuous Rhizobia to simplify inoculation and

2) Increased photosynthate available to nodule and/or improved efficiency of nodule which should produce a large increase in N_2 fixation as described above.

Examples of more speculative long-term impact technologies include:

1) Non-rhizobial N_2-fixing associative symbiosis for non-legume crops,

2) Mycorrhizal associations containing endosymbiotic diazotrophs,

3) N_2-fixing infective but nonpathogenic pathogens, e.g., Agrobacterium, Erwinia,

4) Synthetic N fertilizers by zero direct energy input process,

5) Extension of rhizobial N_2-fixing association to non-legume crops,

6) Transfer of genetic information for N_2-fixing system to crop plants including legumes as well as cereal grains and finally in the most distant future

7) Synthetic gene that codes for small, stable (oxygen and temperature), high turnover, absolute substrate specificity (no H_2 production), zero direct energy requiring (no ATP) N_2-fixing enzyme with appropriate repression by fixed nitrogen.

CONCLUDING THOUGHTS

Some general and specific comments follow regarding the broad topic of this symposium as well as the specific topic of nitrogen input. There is a long-lead time from basic research to implemented applications. In a few selected cases a marriage of basic and applications research is desirable for effective translation. Marriage of basic and applications research in all cases would inhibit some of the usual breakthrough research. Sophisticated basic research usually produces sophisticated (in terms of economics, equipment, and user) technologies requiring high value

in use markets which is not the case in most less-developed countries.
The inherent limited proprietariness of most genetic solutions is
less attractive to the private sector and favors continuous
minimal rather than discontinuous order of magnitude advances.
The drawf wheats and rices of the green revolution may be an exception. The above general comments lead me to suggest that our
expectations should be limited with respect to sophisticated
basic research providing appropriate (in terms of time, technical)
solutions for less-developed countries.

Biological N_2 fixation research has been outstandingly productive of sophisticated basic information during the past 20
years with unfortunately no significant practical applications.
The excessive energy requirement of biological N_2 fixation is a
major and possibly the major limitation. It is recommended that
highly creative basic research on biological and abiological N_2
fixation as well as other N input processes be continued and
encouraged with about an equal emphasis on N_2 fixation and other
N input processes. The world need and the opportunity is so
great as to justify a multifold approach.

REFERENCES

Albrecht, S.I., Maier, R.J., Hanus, F.J., Russell, S.A., Emerich,
     D.W., and Evans, H.J. 1979. Hydrogenase in Rhizobium
     japonicum increases nitrogen fixation by nodulated soybeans.
     Science, 203:1255-1257.
Anon. 1976. 1975 Production Yearbook, Vol. 29, Food and Agri-
     cultural Organization of the United Nations, Rome.
Anon. 1979a. Brighter days ahead in ammonia? Farm Chemicals,
     142(3):13-27.
Anon. 1979b. Southeast Asia faces nitrogen fertilizer glut.
     Chem Eng. News, 57(15):18.
Boerma, A.H. 1979. Food supply of the world. Britannica Book
     of the Year, Encyclopaedia Britannica, Inc., Chicago, pp.150-
     154.
Evans, H.J. and Barber, L.E. 1977. Biological nitrogen fixation
     for food and fiber production: some immediately feasible
     possibilities. Science, 197:332-339.
Garcia, R.L. and Hanway, J.J. 1976. Foliar fertilization of
     soybeans during seed-filling period. Agron. J., 68:653-657.
Giaquinta, R.T. 1977. Possible role of pH gradient and membrane
     ATPase in the loading of sucrose into the sieve tubes.
     Nature, 267:369-370.
Hardy, R.W.F. 1977. Increasing crop productivity: agronomic and
     economic considerations on the role of biological nitrogen
     fixation. In Report on the Public Meeting on Genetic
     Engineering for Nitrogen Fixation. A. Hollaender, ed., U.S.
     Government Printing Office, pp.77-106.

Hardy, R.W.F. 1979. Chemical plant growth regulation in world agriculture. In _Plant Regulation and World Agriculture_, T.K. Scott, ed., Plenum Press, New York, pp.165-206.

Hardy, R.W.F. and Havelka, U.D. 1975. Nitrogen fixation research: a key to world food? _Science_, 188:633-643.

Hardy, R.W.F. and Havelka, U.D. 1977. Possible routes to increase the conversion of solar energy to food and feed by grain legumes and cereal grains (crop production): CO_2 and N_2 fixation, foliar fertilization, and assimilate partitioning. In _Biological Solar Energy Conversion_, A. Mitsui, S. Miyashi, A. San Pietro, and S. Tamura, eds., Academic Press, New York, NY, pp.299-322.

Hardy, R.W.F. and Havelka, U.D. 1978. Research on nitrogen and carbon input to increase domestic crop protein production. In _Protein Resources and Technology_, M. Milner, N.S. Scrimshaw, and D.I.C. Wang, eds., AVI Publishing Co., Inc., Westport, Conn., pp.204-235.

Hardy, R.W.F., Havelka, U.D., and Heytler, P.G. 1979. Nitrogen input with emphasis on N_2 fixation in soybeans. In _World Soybean Research Conference II_, in press.

Heytler, P.G. and Hardy, R.W.F. 1979. Energy requirement for N-fixation by rhizobial nodules in soybeans. _Plant Physiol._, 63S:84.

Huber, D.M., Warren, H.L., Nelson, D.W., and Tsai, C.Y. 1977. Nitrification inhibitors -- new tools for food production. _BioScience_, 27:523-529.

Shanmugan, K.T. and Valentine, R.C. 1975. Molecular biology of nitrogen fixation. _Science_, 187:919-924.

LINKING BASIC RESEARCH TO CROP IMPROVEMENT PROGRAMS FOR THE LESS-DEVELOPED COUNTRIES: BIOLOGICAL CONTROL OF INSECTS

Thomas R. Odhiambo

Director
The International Centre of Insect Physiology and
    Ecology
Nairobi, Kenya

Africa possesses approximately 300,000 known species of insects, although it is widely recognized that many more species still remain to be discovered, characterised, and designated, particularly among the soil inhabitants, the insect parasites and predators, and the insects living on plants of non-economic value to humankind. From this rich and diverse insect fauna, only about 4,000 species - or little more than 1% of the total known fauna - are pestiferous to man and his agricultural produce (Odhiambo, 1975). On a worldwide basis, it is estimated that 3 million insect species comprise the world fauna, of which 3,000 species are agricultural pests and vectors of human and animal diseases; that is, only the minute proportion of 0.1% are pestiferous species (Williams, 1967). Although estimates of this nature might differ in detail, the inescapable fact is that there are a great many insects in this world, and that in the tropical regions one is impressed by their dominant impact on human health, his agriculture, and his very history on this planet (Odhiambo, 1975).

The Genesis has a graphic statement to make on this score:

> "And the evening and the morning were
> the fifth day.
>   And God said, Let the earth bring
> forth the living creature after its kind,
> cattle, and creeping thing, and beast of
> the earth after his kind: and it was so.
>   And God made the beast of the earth
> after his kind, and cattle after their kind,
> and every thing that creepeth upon the earth

after his kind: and God saw that is was good.
     And God said, Let us make man in our
image, after our likeness: and let them have
dominion over the fish of the sea, and over
the fowl of the air, and over the cattle,
and over all the earth, and over every creeping
thing that creepeth upon the earth."

The manner in which man can exercise his dominion over the creeping
things of this world is at the heart of crop protection.

THE SCALE OF THE RURAL PROBLEM

    In a recent study of the food needs of the Developing Market
Economy countries (DMEs), or the Less-Developed Countries (LDCs)
as they were better known, the International Food Policy Research
Institute demonstrated clearly that the major staple food crops
of Asia are largely dominated by the cereals, especially rice,
wheat, and the course grains (such as sorghum and millet), which
together account for 83.4% of the major staples of the region.
A similar situation exists in Latin America, with the cereal staple
crops amounting to 80.0% of the staples, although the composition
is quite different, with a dominant position being taken by the
course grains (Table 1). The situation in Sub-Sahara Africa is a
totally different one; not only do the cereal staples take a
diminished role in forming the food staples in Africa, accounting

Table 1.   Relative Distribution of the Production of Major
           Staples in Developing Market Economy Countries,
           1975-1976

	Asia	Sub-Sahara Africa	Latin America
Cereals	83.4	55.1	80.0
Rice	49.1	5.5	9.6
Wheat	14.7	1.6	15.5
Course Grains	19.6	48.0	54.9
Selected Crops	16.6	44.9	20.0
Root Crops	5.4	30.0	13.1
Pulses	4.7	6.3	5.2
Groundnuts	5.5	8.6	1.7

Source: Food Needs of Developing Countries: Projections of Pro-
          duction and Consumption to 1990," 1977, The International
          Food Policy Research Institute, Research Report 3, Washing-
          ton, D.C.

for only 55.1% of the total staples, but the root crops play a considerable role as a staple food resource (amounting to 30.0% of the staples). Furthermore, 87% of the cereal production in Sub-Sahara Africa is devoted to the course grains (Table 1).

The salient point to be derived from the status report on the food resources of the DMEs is the fact that most basic and production-oriented research has been devoted to a few staples - rice, wheat, barley, and, to a limited extent, sorghum. Course grains as a whole, grain legumes, and root crops have been manifestly neglected in the past. With rice (5.5%) and wheat (1.6%) only comprising a marginal resource base for the food needs of Sub-Sahara Africa (Table 1), it is no wonder that the "green revolution" of Asia (based mainly on rice and wheat) has had almost no ripple effect on Africa.

Discussion of food production almost always leads to a consideration of the rate of increase of food resources as compared to the increase in the human population. It is clear that, when one considers the mean annual growth rates in the period between 1960 and 1975, Latin America showed the most rapid growth, and it registered an output of the major staples at nearly 1.0% above the rate of population increase during the same period (Table 2).

Table 2. Mean Growth Rates of Population and Food Production in Developing Market Economy Countries

	Population		Food Production	
	1960-75	1975-90	Cereals 1960-75	All Major Staples 1975
Asia	2.5	2.5	3.1	2.8
North Africa/ Middle East	2.7	2.9	2.5	2.5
Sub-Sahara Africa	2.6	2.9	1.8	2.2
Latin America	2.8	2.8	3.7	3.6
TOTAL: DMEs	2.6	2.7	2.3	2.9

Source: "Food Needs of Developing Countries: Projections of Production and Consumption to 1990", 1977, The International Food Policy Research Institute, Research Report 3, Washington, D.C.

Table 3.   Annual Rate of Increase in Agricultural Production
            (In Percentage)

Country	Total Agricultural	Food
Chad	-2.1	-2.5
Niger	-1.4	-1.4
Mali	-0.6	-0.9
Upper Volta	1.0	0.6
Central Africa Empire	1.7	1.5
Somalia	2.1	2.1
Uganda	2.3	2.4
Ethiopia	2.6	2.5
Dahomey	2.8	2.0
Tanzania	2.9	3.3
Sudan	5.0	5.0

Source: Biswas, M.R., 1979, Intern. J. Environmental Studies
           13: 207-217 (Based on data from FAO "Production Year-
           book", 1973)

Sub-Sahara Africa, of all the DMEs, performed the most poorly.
All DMEs taken as a whole showed, on average, that the whole
region will become serious food deficit countries - if the trends
almost evident in the 1960-1975 period are allowed to continue
(Table 2). Indeed, if one were to examine the specific situation
in Sub-Sahara Africa, one would find that already some countries
are showing a major deficit in both their food production and their
total agricultural production (Table 3).

    Chad, Niger, and Mali are already experiencing a deficit in
food production; several other countries are having rates of food
production which can hardly keep up with their increasing population
- Upper Volta, Central Africa Empire, Somalia, Ugunda, Ethiopia,
Dahomey, and Tanzania; and only the Sudan has shown an effective
food production increase. Clearly, a major problem of considerable
proportion is brewing up in Sub-Sahara Africa, and it threatens to
be an endemic, long-term problem at that.

    The most optimistic note struck by the governments of the
DMEs this decade, expressed clearly during the World Food Conference
held in Rome five years ago, was their determination to increase
the food resources of their countries by increased reserach and
technology innovation, and by better articulated technology transfer.
Even so, there is still an underlying feeling that many governments,
especially in the DMEs, undervalue the economic contributions of
agriculture - even though over 70% of the population of Sub-Sahara
Africa, for instance, is concentrated in the rural area. Theodore

Schultz (1978) has put the problem in its true perspective:

> "Even though the rural population in low-income
> countries is much the larger, the political market
> strongly favors the urban population at the direct
> expense of rural people. Politically, urban con-
> sumers and industry demand cheap food. Accordingly,
> it is more important politically to provide cheap
> rice in Bangkok than to provide optimum price
> incentives for rice farmers in Thailand."

Incentives are important for the farmer, just as it is vital for
other segments of human endeavour.

One incentive is for the farmer to produce enough of his crops
to sell for other services and goods. A primary incentive for him
is to gather enough food for his own subsistence. I have stated
elsewhere, in considering the question of mixed cropping in
traditional African farming systems, the farmer's primary incentives
(Odhiambo, 1977):

> "He wants to stabilize his agricultural output and
> minimize risks by having mixed cropping system. He
> is not primarily interested in maximizing his yields
> of individual component crops, but he is certainly con-
> cerned with the production of food staples in adequate
> quantities for his family's needs, a variety of vege-
> tables and fruits for their requirements, and a range
> of novelty items such as condiments and plants for
> herbal medication. He is anxious to improve his general
> way of life and enjoy a fuller existence, just as any-
> body else, by having the means for purchasing the
> things he cannot obtain from his own agricultural
> enterprises (such as clothing)."

An approach to the long-range solution of the food resources
of the DME countries in Africa is to take a critical look at the
options available for a rapid increase of the productive activities
within the agro-ecosystem of the region. We know, for example,
that most of the increase in cereal production in all the DME
regions (which amounted to 2.7%) as a whole during the 1960-1975
period was largely due to the increase of output per unit area
(60%); whereas the increase ascribed to the expansion of the arable
land was only 40%. On the other hand, in Sub-Sahara Africa, over
the same period of time, the increase in cereal production was a
mere 1.3%, of which an overwhelming responsible factor was the
expansion of arable land (92%) while that due to the increase of
output per hectare comprised only 8%. If Sub-Sahara Africa is to
meet its total projected 1990 market demand for food production

(all major staples included), then the latter's growth rate would
have to jump from the rate obtained in the 1960-1975 period of 2.2%
per year to one of 4.0-4.4% per year.

This jump in growth rate in food production must presume that
Africa will have solved its problems of a productive agronomic
system, sustained fertility and conservation of agricultural land,
plant protection, considerable reduction (if not elimination) of
post-harvest losses, and other practices necessary to pull Africa
through to a system of higher, sustained crop yields. The first
three problem areas require a great deal of basic research to the
specific questions of tropical agriculture. The fourth problem
seems to be simply a technological one, as a U.S. National Academy
of Sciences report (1978) has stated:

> "Our study confirms that there is no known simple,
> inexpensive technology that can, by itself, make a
> profound impact on post-harvest losses. On the con-
> trary, post-harvest food conservation can be achieved
> only through a combination of location-specific
> organization, problem identification, training, infor-
> mation, and adapted technology." (p5).

The trouble is that the major sources of agricultural losses
in the rural areas are not really in storage and during processing
- although these are quite large by themselves - but on the farm,
during the growth and developmental stages of the crop. Crop
protection for a monoculture system of agricultural production is
a very different proposition from that of a mixed cropping system,
which is the predominant system for food production in rural Africa.

MIXED CROPPING ENTOMOLOGY

It is thought that there are something like 250,000 species
of flowering plants known in the world. Of this number, less than
20 species have been domesticated and used regularly by humankind
as food staples (Wilhelm, 1976):

- Six cereals - wheat, rice, rye, maize, sorghum, and millet
- Four legumes - bean, soybean, chick pea, and lucerne (or
  alfalfa)
- Four root crops - Irish potato, sweet potato, cassava, and
  taro
- Two tree crops - banana and coconut
- Two sugar crops - sugarcane and beet.

These few plant species, cultivated as they are in pure stands
quite often, makes the crop component of the total agroecosystem

greatly simplified in terms of the known natural plant communities.
This simple, artificial plant community has been tremendously
successful in giving the modern farmer a means to increase his
agricultural productivity through mechanization, heavy fertilizer
applications, use of pesticides, and adoption of specifically
developed high-yielding crop varieties of known architecture.

The same cannot be said for the highly complex cropping system
most widespread in the tropical areas of Africa and Asia. Indeed,
in a rather pointed summary, Wilhelm (1976) had this to say:

"Yet we cannot do as primitive cultures did and still
do, except in our home gardens, namely grow a jumble
of different food plants together in small plots, or
even as the Aztecs did, plant corn and beans together,
the beans climbing the corn. For the sake of efficiency
of all operations and for the essentiality of high
yields modern agriculture is built around genetic
uniformity, monocrop culture, and machine harvesting."

The jumble of crops in small farm holdings in Sub-Sahara
Africa is a characteristic feature of the African traditional farm
systems (Okigbo, 1978). Farm sizes are small, of about 2-5 hectares;
and in these farms are grown together a great variety of food
plants in multiple intercropping patterns, in relay cropping systems, and in patch intercropping systems of annuals, perennials, or
both. For instance, in a compound (or homestead) garden in the
forest zone of West Africa, more than 20 different crop species
may be grown together representing staple food species, oil crops,
condiments and spices, masticants and stimulants, plants for herbal
drugs, fibre plants, plants for firewood, plants to provide animal
feed, shade trees and ornamentals, boundary plants, and plants
for religious and social functions. This system of mixed cropping
is extended to field crops. The wide range of data provided by
Okigbo (1978) shows quite clearly that the practice is prevalent
throughout Africa. Even with high-value food crops such as maize
and soybean, mixed cropping is the rule rather than the exception
(Table 4).

Mixed cropping looks messy. It is certainly not amenable to
present-day agricultural machinery, which has been specifically
developed for the monoculture system of cropping.

But there is some evidence that the traditional intercropping
system gives higher total yields than those given by the same crops
grown in monoculture (Odhiambo, 1977; Okigbo, 1978). This is
apparently due to three principal factors that characterize the
mixed cropping agronomic system (Harwood, 1976; Okigbo, 1978):

Table 4.   Level of Mixed Cropping

| | Percent Mixed Cropping | |
	Nigeria	Uganda
Maize	75.5	83.7
Millet	89.6	47.6
Cowpeas	99.0	61.8
Groundnut	95.5	56.3
Soybean	100.0	–
Cocoyam	86.4	–
Banana	–	42.1
Sweet potato	–	13.7
Cassava	26.8	49.9
Cotton	80.1	26.2

Source: Bede N. Okigbo, 1978, "Cropping Systems and Related
          Research in Africa," Association for the Advancement
          of Agricultural Sciences in Africa, Addis Ababa.

· Losses due to pests and diseases are minimized, and diseases
  and pests appear to spread more slowly
· Losses due to adverse environmental conditions are less
· Soil is better protected against insolation and erosion.

Okigbo (1978) quotes examples in Sub-Sahara Africa, where intercropping (cereal/legume, legume/root crop, or other crop mixtures)
gave higher total yield per unit area than when the same component
crops were grown as sole crops, even when these were grown under
improved levels of technology. This was apparently a result of
a more efficient system for the utilization of ecological resources,
plants of different heights, rooting systems and nutrient requirements exploiting the same unit area. Harwood (1976) reports somewhat similar results in South-East Asia, where he has shown that in
a maize/groundnut intercropping system, there is a dramatic drop
in the incidence of downy mildew and fewer maize stem-borers than
under monocrop conditions (Table 5).

This type of evidence has not yet reached overwhelming proportions, although we can now begin to state that pests under
tropical agronomic conditions can buld up to epidemic proportions
over the "green revolution" areas of South-East Asia. The large
areas of monocrop rice are fragile agroecosystems, especially
when single varieties of crops are predominant. It can also be
stated that we need to know a great deal more about mixed cropping
– using modern tools of research in a systematic manner – in order
to understand thoroughly the system as a whole, its peculiar

Table 5. Comparison of Maize Stem-Borers on Maize With or Without
          Groundnut Intercropping (Philippines, 1972 Wet Season,
          at 56 Days after Planting)

Maize Population (Plants/Hectare)	40,000	40,000	20,000	20,000
Groundnut Intercrop	yes	no	yes	no
Border-infested Plants (%)	30	86	21	60
Borers (Insects/ Plant)	0.4	2.6	0.4	2.0
Feeding or Exit Holes (Holes/Plant)	0.5	3.1	0.5	2.1
Feeding Tunnels (Tunnels/Plant)	0.4	2.7	0.5	2.0

Source: Richard R. Harwood, 1976, in: "Nutrition and Agricultural
          Development", N.S. Scrimshaw and M. Behar, eds. Plenum
          Press, New York.

advantages, and how to make it a more efficient alternative,
especially for the small farmer.

     It is obvious that research on mixed cropping has been grossly
neglected. Experimentation is certainly more complex, it is more
difficult to apply the present-day inputs (pesticides, fertilizers,
and others), and mechanization for planting and harvesting has
not been designed with mixed cropping systems in mind. The
diversity of food crops within a cropping unit itself is a dis-
couragement to most agricultural scientists, who are used to an
analytical approach dealing with a few components in a simple
agroecosystem. This is one reason why the green revolution started
in South-East Asia, and later in Latin America. Both regions
live mainly on rice and wheat, and they have not adopted a traditional mixed cropping system of such a diverse nature as that
characteristic of the small farmer in Sub-Sahara Africa. If there
is going to be a green revolution in Africa, it is most likely
going to take an evolutionary path.

     With the small farmer still dominant in Africa, it is a matter
of concern that research on mixed cropping has been so slow in
reaching a priority position. One can state quite confidently
at this time that "mixed cropping entomology" does not exist.

     Our entire plant protection practice in Africa - as it is in
other regions of the world - is based solidly on the entomology of
single crops, or on the biology of single pest species. This
approach is evidently appropriate for the temperate regions, where

monocrop systems are the prevailing tradition, and where flora and
fauna are poor in species diversity. In the tropics, particularly in tropical Africa, where mixed cropping covers much the
largest area under agriculture, using "monocrop entomology" to solve
pest problems arising from mixed cropping practices is not only
inappropriate and irrelevant, but indicates our scientific blindness
to the situation so apparent in our target problem area. We simply
have to direct our minds and scientific talent to long-range basic
research on "mixed cropping entomology", to discover the principles
governing pest biology under mixed cropping systems, to understand
the plant's own system of defense under these situations, and to
begin to apply this knowledge to the design of appropriate plant
protection practices for this particular agroecosystem.

We can begin on this long road to gathering a corpus of knowledge of mixed cropping entomology by examining a basic standby of
traditional plant protection - that of crop resistance.

CONVENTIONAL ATTITUDE TO CROP RESISTANCE

Except under unusually favourable conditions, where you can
fight insect pests by eating them (Odhiambo, 1978), there are six
major approaches to insect control:

- The use of insecticides, alone or as part of an integrated
  system of pest management
- The sterile insect release method (SIRM), first celebrated by
  the success of E.F. Knipling in attenuating the screw-worm
  menace in southern USA by the release of sterile-male flies
- Biochemical control by the adoption of several avenues, still
  mostly in the research and development stage; for instance
  pheromones, hormones, and antifeedants
- Cultural and other ecological techniques; for instance, the
  use of crop rotation as a sanitation method for pest control
- Biological control, by the action of parasites, predators,
  and pathogens of the pests themselves
- Intrinsic resistance factors in plant hosts against their
  associated pests.

The last two approaches to insect control are under-rated and
under-researched, especially among the developed world of Eurpoe
and North America (Huffaker, 1971). They have been wholly over-
shadowed by the almost complete faith in the efficacy of the
chemical insecticidal weapon. As Doutt and Smith (1971) have so
well put it, the over-dependence on insecticides has "led to a
closed, circular, self-perpetuating system with a completely
unilateral method of pest control. Alternative measures were not
explored or, if known, they were ignored. The chemical cart ran
away with the biological horse."

The mixed cropping system is much too complex to rely on the simplistic model of the use of insecticides for pest control. The plant host/insect relations, and the ecology and bionomics of the mixed cropping system, which must be the basis of crop protection, are much too complicated to be solved by the simple expedient of using the insecticide blunderbuss. The development of resistance to insecticides, the development of cross-resistance to several classes of chemical insecticides, the impact on non-target organisms (including predators and parasites of the pests themselves), the development of secondary pests as a result of the elimination of natural biological agents for the current pests, our ignorance of the precise ecological situation in mixed cropping, and the high cost of chemical pesticides to the small rural farmer, are all factors that militate against the use of pesticides as the main plank for pest control under mixed cropping systems in Africa.

Under mixed cropping, the system of pest management may appropriately be considered as one of containment, rather than one of eradication. The "sacred insecticidal cow" has no pedestal to stand on in the rural mixed cropping system. Crop resistance has a major role to play. One principal consideration is the economics of the practice. It has been calculated that the cost of developing resistant wheats for the control of four major pests of wheat (Hessian fly, stem sawfly, alfalfa aphid, and maize borer) amounted to 462 professional man-years and U.S. $19.3 million invested in research (Dethier, 1976). The wheat farming community was, as a result, saved $308 million a year. Since a resistant wheat variety may be effective for about 10 years, the accumulated savings for the farming community over this period was $3 billion, a return of 300:1 on the research dollar originally invested on the development of the pest-resistant wheats. By any measure of economic returns, this is an impressive performance, even if it were to be proved to be one magnitude less in effectiveness.

In spite of this impressive performance, North America has not used crop resistance as a significant approach to insect control. However, some useful screening of crops for resistance to insect pests has been reported during the period 1966 to 1977 (Kennedy, 1978). Among the more than 30 crops screened, 5 fruit crops demonstrated resistance to 6 associated insect pests:

- Citrus - Aonidiella auranti
- Pear - Psylla pyricola
- Raspberry - Amphorophora agathonica
- Strawberry - Chaetosiphon fragaefolii, Tetranychus urticae
         (two-spotted spider mite)
- Filberts - Phycoptella avellanae.

Among vegetable crops, the heritability of resistance was demonstrated for 30 insect species in 11 crops. It seems reasonable to

suppose that, in a mixed cropping agroecosystem, where the plant mixture is often quite complex, insect resistance offered by individual crop components of the mixture to the multitude of insect
species, is a much more rational ecological approach to pest management than the chemical-pesticide route. In DMEs, where the farmer's
crops are usually of high significance for subsistence but of low
value in the open market, where crop husbandry is normally carried
out in small scattered plots, and where the farmer meets especially
difficult economic constraints, crop resistance offers the first
line of attack for crop protection.

There are other advantages to varietal resistance as a tool
for crop protection. Varietal resistance is effective to even low
pest populations (whereas the use of insecticides can only be justi-
fied when pest populations reach the economic injury level); and
it is thought that the cultivation of resistant varieties leads to
the emergence of a population of restless pests, which are then
exposed over a longer period to the depredations of parasites and
predators. This is not to say that there are no problems attendant
on the use of varietal resistance for insect control. One immediate
problem is that we are largely ignorant of the mechanisms operating
in a resistance situation. The other is the development of new
biotypes of insect pest species able to attack previously resistant
crop varieties.

A pioneering researcher in crop resistance to insect attack,
Painter (1951), classified the main means for resistance among crop
plants as: "non-preference" (in which the host plant becomes un-
attractive to the insect pest for the purposes of feeding, oviposi-
tion, and shelter); "antibiosis" (in which the host plant has an
adverse effect on the growth and survival of the insect and its
progeny); and "tolerance" (in which the host plant can withstand
heavy insect attack but is still capable of producing a good crop).
The various chemical or physical mechanisms plants have employed
to render themselves resistant to insect attack are diverse (Table
6), and we are now only beginning to explore this field with some
vigour and rigour. A good example is the research on the resistance of Curcubitacae to the cucumber beetle. Some Cucurbitacae
are resistant to the beetle because of absence of cucurbitacins (a
class of tetracyclic triterpenoids). The cucurbitacins have a
bitter taste to humans, but act as feeding stimulants for the
beetles (Kennedy, 1978).

The resistance phenomena we are considering here are those
that are inheritable. Some resistance is of a polygenic nature.
For instance, strawberry resistance to the spider mite (Tetranychus
urticae) is controlled by multiple genes; breeding for resistance
is therefore carried out by adopting a recombinant strategy - which
involves repeated crossing and selfing of the most resistant plants,

Table 6.   Examples of the Biochemical Nature of Resistance of
           Host Plants to Pests

Host plant	Chemical Conferring Resistance	Insect pest	Effect on Insect
Alfalfa	trichome exudate; high saponin content	alfalfa weevil; white grub	antibiosis non preference, antibiosis
	medicagenic acid	potato aphid potato leaf-hopper	antibiosis
Apple	phenolics (gallic, tannic, coumaric acids; quercetin, naringenin and catechin	apple maggot	antibiosis
Barley	benzyl alcohol	greenbug	antibiosis
Chestnut	high catechol tannin content	chestnut gall wasp	antibiosis
Maize	high DIMBOA content;	European corn borer;	antibiosis
	unknown factor in silk	corn earworm	antibiosis
Cotton	high gossypol content	ca.10 different insect pests	toxicity
Cucumber	absence of cucurbitacin	cucumber beetle	nonpreference
	presence of cucurbitacin	two-spotted spider mite	nonpreference
Rice	low asparagine content; unknown chemical(s)	brown planthopper brown planthopper	nonpreference failure of egg eclosion
	high silica content;	striped borer, yellow borer	antibiosis
	benzoic and salicylic acid;	striped borer	toxic
	steam distillate extract	striped borer	nonpreference antibiosis
	lower nitrogen and starch	yellow stem borer	nonpreference
Strawberry	essential oils	two-spotted spider mite, strawberry spider mite	repellence
Sweet clover	ammonium nitrate	sweet clover weevil	feeding deterrence
Tobacco	trichome exudate	bird-cherry-oat green peach aphid, tobacco hornworm	toxic
Tomato	unknown chemical(s)	carmine spider mite, two spotted spider mite	repellence

Source: Pathak, M.D. and Saxena, R.C., 1976, Current Adv. Plant Sci.
          8: 1233-1252.

with little or no selection for resistance in the early generations.
It is considered that the more useful and long-lasting crop resistances are those conferred by polygenic mechanisms. For instance, it
is reputed that a variety of apple ("Winter Majestin") produced as
a resistant variety to woolly apple aphid in 1831 is still resistant.
Some varieties of maize recognized in 1949 as resistant to the
European corn-borer, Ostrinia nubilalis, are still resistant after
continuous cultivation since then. Both sources of resistance are
thought to be controlled by multiple genes (Dethier, 1976).

Some other resistance is under monogenic control; here the
resistance is simple, and the breeding strategy is equally uncom-
plicated, consisting of a simple backcross breeding program with
commercial-type recurrent parent. A good example is the resistance of red raspberry to the aphid, Amphorophora agathonica. The
major problem with this simple mechanism of crop resistance is that
it is fairly easily overcome by the development of resistance-
breaking pest biotypes. A dramatic example has been provided by
the recent history of the rice brown planthopper, Nilaparvata lugens
(Stal), a delphacid bug which was only a minor pest in South-East
Asia, but in recent years has become an epidemic pest following the
introduction and wide cultivation of green revolution rice varieties
in the region. Since mid-1977, the International Centre of Insect
Physiology and Ecology (ICIPE), in Kenya, and the International Rice
Research Institute (IRRI), in the Philippines, have been collaborating
on the development of new strategies for controlling the brown
planthopper. The program involves breeding high-yielding rices with
good agronomic characteristics and insect resistance factors (Pathak
and Saxena, 1978; Saxena, 1978).

A notable feature of the rice brown planthopper (BPH) problem
is that the resistances that have so far been identified have a
narrow genetic base, being of the monogenetic type, and this has
led to the rapid development of resistance-breaking BPH biotypes
throughout the South-East Asia region. From 1972, BPH has plagued
fields planted with high-yielding, high-tillering rice varieties
grown in the Philippines under heavy applications of fertilizers.
This micro-environment seemed to especially favour BPH and populations shot up in many areas in the Philippines. It was not long
before the first BPH biotype was discovered (Heinrichs et al, 1978).
Since then, a number of other biotypes have emerged, to bypass the
resistance of other widely cultivated rice-resistant varieties.
By the end of 1977, the estimated losses in 11 countries of tropical
Asia due to BPH epidemics had reached U.S. $300 million; in Indonesia
alone the losses were put at $100 million. A factor in BPH resur-
gence in the Philippines, Indonesia, India, Bangladesh, and other
Asian countries is due, so it is believed, to factors induced by
the use of insecticides. Natural enemies that previously kept
BPH in check have been eliminated by insecticides, and where sub-
lethal rates of insecticides applications have been made, BPH

feeding has been greatly enhanced, reproductive rate has therefore increased, and nymphal duration has been shortened as soon as insecticide toxicity is lost.

Four biotypes of BPH are known in the Philippines (and a few more have also been described from other South-East Asian countries). These biotypes demonstrate four resistance genes:

- Rice varieties resistant to BPH biotype 1 carry the dominant Bph1 resistance gene (found in the resistant rice variety 'Mudgo')
- Rice varieties resistant to BPH biotype 2 carry the recessive bph2 resistance gene (found in the resistant rice variety ASD7)
- Resistance genes Bph3 and bph4, which show resistance to all known BPH biotypes, have been identified recently in some tall traditional rice varieties at IRRI.

As most rice varieties in the Philippines (and in other tropical Asian countries) have a narrow, monogenetic base for resistance to BPH, their stability is vulnerable and some thought is being given to the slowing down of the consequences of this vulnerability by the sequential release of such resistant varieties of rice (Heinrichs et al, 1978).

One may assume that the most readily available source of resistance should be the aboriginal home of the crop concerned. As a matter of fact, this does not seem to be the experience of plant resistance hunters so far. For instance, out of 866 collections of the African rice, Oryza glaberrima Steud. in the IRRI rice germplasm bank, 300 were found to be highly resistant in the Philippines to the green leafhopper, Nephotettix virescens (Distant), which is not known to occur in Africa at all, and which is a major pest of the Asian rice, Oryza sativa L. (Pathak and Saxena, 1978). The strategy for the screening of crop plants for resistance should therefore be to seek resistance among rice strains grown in areas indigenous to the pest, to explore for resistance in closely related species of rice (and there are about 20 species of rice in the world, only one of which is predominant and is the preferred commercial rice, the Asia rice and having 3 ecogeographic races, and one other cultivated rice species, the African rice occupying only 4% of the world rice hectarage), and to screen for the entire rice germplasm (now comprising some 100,000 strains) for the fortuitous recognition of resistant cultivars (Kennedy, 1978; Pathak and Saxena, 1978).

We still know rather little about the actual mechanism of rice resistance to BPH biotypes. As a first attempt at elucidating the biochemical mechanism involved, Saxena (1978) has recently shown that in biotype 1, asparagine and valine were the most phagostimu-

latory amino-acids at 4% concentration. In biotype 2, alanine was
the most phagostimulatory, while in biotype 3, valine and serine
gave the most phagostimulation. Saxena related these amino-acid
phagostimulation responses to the levels of amino-acids found in the
various resistant rice varieties. Obviously, there is still a long
way to go before we can have a full understanding of the biochemical and biophysical bases of resistance to BPH attack. We are even
more ignorant of how biotypes arise — yet these two sources of information must be the foundation upon which a large part of our
future crop protection strategy for the rural farmer should be based.

There is no question that plant resistance as a method of pest
management will have to become a highly sophisticated system of crop
protection, founded on basic research, and linked to a highly articulated technological service. Included should be a central breeding
station, a central registration mechanism for crop cultivars with
different pest resistance characteristics (and other desirable
agronomic characteristics), a monitoring and surveillance service
for the major pests and their biotypes, and the means to produce
and distribute the right seeds to the target agronomic area. In
this manner, the small farmer will participate in protecting his
crops — even in the mixed cropping system — with the aid of the
state and with little direct expense from his own meagre resources.

Table 7. Allelochemic Factors in Plants and Their Impact on the
Corresponding Behaviour and Physiological Responses of
Attendant Insects

Allelochemic Factor	Effect on Insect Response
ALLOMONES	GIVE ADAPTIVE ADVANTAGE TO HOSTPLANTS
Repellents	Orient insect away from plant
Locomotor stimulants	Increase walking or flight activity
Suppressants	Inhibit biting or piercing
Feeding deterrents	Decrease duration of feeding
Oviposition inhibitors	Interrupt egg-laying
Metabolic inhibitors	Disrupt normal physiological pro-cesses
KAIROMONES	GIVE ADAPTIVE ADVANTAGES TO INSECT PESTS
Attractants	Orient insect towards plants
Arrestants	Slow or stop movement
Excitants	Evoke biting or piercing
Feeding stimulants	Increase duration and rate of feeding
Oviposition inducers	Induce egg-laying
Metabolic regulators	Regulate normal life processes

Source: M. Kogan, 1975, "Introduction to Insect Pest Management",
John Wiley & Sons, New York.

INSECTS AS CHEMISTS

Insects are by far the most accomplished chemists among the terrestrial animal groups on this planet. Most of the external integument is one big chemosensory observatory, and the lining of the internal integument is similarly punctuated by a galaxy of chemo-analytical structures. Their social behaviour is largely regulated by a chemical language that we are only beginning to unravel; their response to their hosts is considerably influenced by the allelochemic factors within or surrounding the relevant hosts (Table 7). They have learned to use the biochemicals naturally occurring in their hosts for their own peculiar uses. A germane illustration is farnesol, a sesquiterpenoid synthesized by plants and located mainly in mesophyll cells of leaves.

An important mechanism for ensuring that the rate of plant transpiration is regulated to a level which can be supported by water uptake is that of stomatal movement. Mansfield and his colleagues (1978) have recently shown that abscisic acid, another sesquiterpenoid, is released from mesophyll chloroplasts into the guard cells of the stomata under water stress conditions, thus closing the stomata. The interesting new finding is that it is all trans farnesol which is responsible for the alteration of the permeability of chloroplast membranes, thus permitting the release of abscisic acid from the chloroplasts in the mesophyll. An amazing fact is that the juveline hormones (JHs) of insects are based on this 15-carbon skeleton, with certain activity-optimizing modifications which make the JHs possess both juvenilizing and gonadropic activity (Siddall, 1977). It is possible that plants have already, in being, an anti-JH or a precursor anti-JH biochemical which abolishes the activity of the anti-transpirant farnesol?

By a remarkable parallel argument, Bowers et al (1976) have isolated from the common bedding plant, Ageratum houstonianum, two compounds with anti-JH activity, precocene-1 and precocene-2 (which is 6,7-dimethoxy-2, 2-dimethyl-chromene), the latter being by far the more active compound. Precocene is probably not the anti-JH we need for the majority of plant protection work as it is active only on a restricted number of insect species (mostly Hemiptera). We have to still identify the anti-JH biochemical for the holo-metabolous insects, which comprise the vast majority of tropical crop pests. But the main point of this illustrative story is that we should now meticulously follow the chemical route of insects back to their host plants, in order to enable us to glean clues as to their sources of chemical bricks and the plant's own crucible of anti-bricks.

THE FUTURE OF CROP PROTECTION IN THE DEVELOPING COUNTRIES

The intention of this address is not to give a blueprint for future fundamental research needed to resolve the plant protection problems of the DMEs, nor is it to give updated technical recipes for the application of the principles of plant protection. Rather, it is to bring out much more clearly than it has been done before two salient perspectives critical to the pest management problems of the DMEs, particularly in Sub-Sahara Africa:

- Firstly, food production in Africa consists of an undervalued complex of component crops planted in predominantly small plots of land in a relatively sparsely populated region (averaging 20 inhabitants per square kilometre). This requires a different type of input of scientific research and technological innovation to strengthen the traditional bases on which agriculture has existed over thousands of years.
- Secondly, it is paramount that we understand the mixed cropping entomology of the traditional farming system, which has a great deal of ecological reasons to make it the effective choice of the small farmer in the tropics. Mixed cropping entomology is a cornerstone of our future strategy for plant protection in the rural farming areas of the great majority of the tropical peoples of Africa.

It has not been possible to go into great lengths to sketch out what areas of concern we should tackle, although it has seemed practical to consider crop resistance to insect depredations as the first line of attack in this new endeavour.

I believe that, as the story of pest biology under mixed cropping systems unfolds, it is going to become more and more obvious that our approach to insect pests must be one of containment, of sharing what we have with the pests - unequally, with humankind arranging matters in such a way that they get away with the larger portion of the spoils - but always keeping in the forefront of our minds that if we are to give ourselves an equal opportunity to fight the insects, then we must become as good biochemists as the insects already are. Otherwise, it will continue to be an unequal fight - with the insects winning the battles, although we tend to win the skirmishes.

REFERENCES

Anonymous, 1977, "Food Needs of Developing Countries: Projections of Production and Consumption to 1990", The International Food Policy Research Institute, Research Report 3, Washington, D.C.

Biswas, M.R., 1979, Nutrition and agricultural development in Africa, Int. J. Environ. Stud., 13: 207.

Bowers, W.S., Ohta, T., Cleere, J. and Marsella, P., 1976, Discovery of anti-juvenile hormones in plants, Science, 193: 542.

Dethier, V.G., 1976, "Man's Plague: Insects and Agriculture", The Darwin Press, Princeton.

Doutt, R.L. and Smith, R.F., 1971, The pesticide syndrome - diagnosis and suggested prophylaxis, in: "Biological Control", C.B. Huffaker, ed., Plenum Press, New York.

Harwood, R.R., 1976, The application of science and technology to long-range solutions: multiple cropping potentials, in: "Nutrition and Agricultural Development," N.S. Scrimshaw and M. B'Ehar, eds., Plenum Press, New York.

Heinrichs, E.A., Saxena, R.C., and Chelliah, S., 1978, Development and implementation of insect pest management systems for rice in tropical Asia, in: Seminar on "Sensible Use of Pesticides", sponsored by ASPAC Food and Fertilizer Technology Center, Tokyo (Mimeographed).

Huffaker, C.B., 1971, Preface, in: "Biological Control", C.B. Huffaker, Huffaker, ed., Plenum Press, New York.

Kennedy, G.G., 1978, Recent advances in insect resistance of vegetables and fruit crops in North America: 1966-1977, Bull. Entomol. Soc. Am., 24: 375.

Kogan, M., 1975, Plant resistance in pest management, in: "Introduction to Insect Pest Management," R.L. Metcalf and W.H. Luckmann, eds., John Wiley & Sons, New York.

Mansfield, T.A., Wellburn, A.R. and Moreira, T.J.S., 1978, The role of abscisic acid and farnesol in the alleviation of water stress, Philos. Trans. R. Soc. London, Ser. B, 284: 471.

National Academy of Sciences, 1978, "Postharvest Food Losses in Developing Countries", Board of Science and Technology for International Development, Washington, D.C.

Painter, R.H., 1951, "Insect Resistance in Crop Plants", Macmillan, London.

Pathak, M.D. and Saxena, R.C., 1976, Insect resistance in crop plants, Curr. Adv. Plant Sci., 8: 1233-1252.

Pathak, M.D. and Saxena, R.C., 1978, "Breeding Approaches in Rice", IRRI, Los Banos (Manuscript).

Odhiambo, T.R., 1975, "This is a Dudu World", The International Centre of Insect Physiology and Ecology, Nairobi.

Odhiambo, T.R., 1977, "Science and Technology for the Rural Farmer", The International Centre of Insect Physiology and Ecology, Nairobi.

Odhiambo, T.R., 1978, "The Use and Non-Use of Insects", The International Centre of Insect Physiology and Ecology, Nairobi.

Okigbo, B.N., 1978, "Cropping Systems and Realted Research in Africa", Association for the Advancement of Agricultural Sciences in Africa, Addis Ababa.

Saxena, R.C., 1978, "Biochemical Basis of Insect Resistance in Crop Plants", IRRI, Los Banos (Manuscript).

Schultz, T.W., 1978, Constraints on agricultural production, <u>in</u>:
    "Distortions of Agricultural Incentives", T.W. Schultz, ed.,
    Indiana University Press, Bloomington.
Siddall, J.D., 1977, Juvenile hormones and their analogs, <u>in</u>:
    "Study Week on Natural Products and the Protection of Plants",
    G.B. Marini-Bettolo, ed., Pontificiae Academiae Scientiarum
    Scripta Varia, Rome.
Wilhelm, S., 1976, The agroecosystem: a simplified plant community,
    <u>in</u>: "Integrated Pest Management", J.L. Apple and R.F. Smith,
    eds., Plenum Press, New York.
Williams, C.M., 1967, Third-generation pesticides, <u>Sci. Am.</u>, 217:
    13.

USE OF PREDATORS AND PARASITOIDS IN BIOLOGICAL CONTROL

C.B. Huffaker

Division of Biological Control
Department of Entomological Sciences
University of California
Berkeley, California 94720

I.   INTRODUCTION

First, let me say that some of the work with parasitoids,
predators and pathogens in biological control is only _suggestive_
of utility in developed or developing countries. For other work
we are reporting on what is being done _now_, some of it in developing
countries. Much of the "classical" biological control by introductions of natural enemies has occurred in developing countries.
Such biological control is often little dependent on changing
traditional modes in agriculture; it is in fact especially
threatened by changing the mode to the "modern", i.e., adoption
of heavy dependence on disruptive pesticides. Moreover, natural
enemy augmentations seem particularly relevant to countries where
low-cost labor or channelling of labor prevails.

The premise on which biological control rests is that in
certain circumstances many species are held at low endemic densities
by their natural enemies, that is, parasites, predators and pathogens. For many species that are pests or potential pests, their
natural enemies or natural enemies of their close relatives may
provide a control solution. The premise is itself only one aspect
of the concept of 'balance of nature,' which means that populations
are restricted in the numbers they can attain. As a species
increases in density, it uses up needed resources, defiles the
place in which it lives, and attracts an increased intensity of
inimical factors such as predators, parasites and disease.

There is much evidence of the profound role that pathogens,
parasites and predators play in this game of balance in nature.

173

The awesome and recurrent effects of human disease on human populations throughout history and throughout the world, prior to the
great advances in modern medicine, suggest the potential role of
pathogens. The role of predators and parasitoids in modern applied
control of insect pests was first demonstrated spectacularly with
the introduction of the Vedalia beetle, Rodolia cardinalis (Muls.),
from Australia into California in 1888 to control cottony cushion
scale, Icerya purchasi Mask., which had invaded and was destroying
the burgeoning young citrus industry. So successful was this introduction that the control has remained virtually complete to this
day despite heavy use of various insecticides. Only the use of
DDT in the late 1940's proved to have a major disruptive role, and
its use since then has been avoided.

The potential impact of predators and parasitoids has since
been broadly demonstrated by other introductions but, just as
significantly, the powerful role of native natural enemies in
holding native potential pest insects under control was demonstrated most spectacularly with the advent and wide use of DDT in
the late 1940's and early 1950's. Just as DDT, when used on citrus,
killed the vedalia beetles and caused explosive resurgences of the
cottony cushion scale, uses of DDT and of other broad spectrum
synthetic insecticides so interfered with various natural enemies
that the populations of both alien and native species that had
formerly been of relatively little or no economic consequence
increased rapidly. This biological control of native "non-pest"
species by native natural enemies exists all around us, and without
it our pest problems would be much greater than they are. This
unseen biological control has been likened to the hidden part of the
iceberg, about 80 to 90 percent of the total. This use of disturbing
pesticides simply uncovered the existence of natural controls (e.g.,
DeBach 1974), and has created many "induced pests."

Examples of pesticide-induced pests include the classical
build-up of spider mites all over the world (Porter 1947, McMurtry
et al. 1970 and Huffaker et al. 1970), of bagworms in Malaysia
(Wood 1971), bollworms and budworms in cotton (Adkisson 1971),
cyclamen mites in strawberries (Huffaker and Kennett 1956), and
a horde of others (Ripper 1956). A recent notorious example is
that of the tobacco budworm in the Rio Grande areas of Mexico and
Texas (Adkisson 1971) and in California (Smith and Reynolds 1977).
Treatments of cotton for boll weevil in the Rio Grande area of
Mexico and Texas had destroyed the formerly effective natural enemy
control of the bollworm and tobacco budworm, which led to their
becoming pests, and this was followed in 1971 by selection of a
biotype of the budworm that was 170 or more times as resistant
as formerly to insecticides used against it. More recently, in
the Imperial Valley of California, the same pattern followed from
treatments intended to "eradicate" the pink bollworm. In both

cases, as many as 20 or even 30 treatments per season failed to control the pest. Relief in Texas was obtained by changing the whole strategy of control of boll weevils, allowing recovery of biological control of the budworm (see below); however, in the Rio Grande area of Mexico, cotton production was abandoned (Adkisson 1971). In California no solution has yet been developed, and the flood of budworms into vegetables and lettuce is even more threatening because growers are still trying ineffectively to control the pest with insecticides. Yet, Smith and Reynolds (1977) state categorically: "In our considered view, the integrated pest management approach has literally rescued cotton production from economic disaster in the few places where its intrinsic value has been recognized, e.g., Peru and Texas." They add, relative to the Imperial Valley situation, "When will we ever learn?"

Classical biological control is simply an effort to establish in a new area a biological control link that has long existed in another area. This is accomplished by discovering an effective natural enemy from the former native home area of the pest species and introducing it into the region that the pest has invaded. It is also possible to conserve and to augment the numbers and effectiveness of natural enemies by eliminating use of pesticides or using them in less inimical ways, and by altering the environment or agricultural practices to favor them. Augmentation may also be achieved by insectary production and releases into the fields, and by other means (see below).

These basic concepts, involving the roles of parasites and predators in controlling and regulating their hosts' populations, the theoretical implications of their complex characters, adaptations and behaviors in these roles and the empirical record itself, will be treated here in brief summary, in the hope of conveying an impression of the importance of biological control. This discussion is concerned only with true predators and patasitoids. The important role of pathogens is not within the assigned scope of this paper.

II.   THE EMPIRICAL BASIS OF BIOLOGICAL CONTROL

Following the first spectacular classical biological control success, a world traffic in exploration and importation of biological control agents developed. Classical biological control, using parasitoids and predators, has resulted in some 157 cases of substantial or complete success with particular insect pest species. With the duplication of these successes in more than one major geographic area (continent or island group), and there have been at least 246 substantial to complete successes (Laing and Hamai 1976). Likewise, there have been some 39 cases of substantial to complete success in the control of weed species in distinct major geographic areas, mainly by use of insect predators (Table 1).

Table 1. Ratings of Biological Control of Weeds: Successes
          Listed for Every Major Geographic Area Into Which
          Importation has Occurred. (From Laing and Hamai 1976).

Degree of Success	Islands	Continents	Total
Complete	7	6	13
Substantial	9	17	26
Partial	4	14	18
Total	20	37	57

In the practical and political sense, it is upon this monu-
mentally successful empirical base that the practice of biological
control rests. At every stage in development of this discipline
funds to support it and enthusiasm to pursue it have derived from
some new, surprising or spectacular practical success. These
successes must in large measure be termed empirically derived,
because no clear explicit understanding of the detailed ecological
behavior and performance of the agents considered for introduction
formed the basis of making specific introductions. However, in
many instances the observed rarity of the pest in the native home
area where the natural enemies were collected, and its devastating
status in the target area which lacked such enemies, suggested the
theoretical possibility of a successful introduction. This sug-
gestion is in line with general theory of the role of natural
enemies in the balance of nature---that they operate as density
dependent regulating factors. It does not explain how they achieve
such a role. It was presumed that if well adapted, they would
thrive and if poorly adapted, they would not, and that by introducing a complex of such predators, parasitoids and pathogens, one
or two (perhaps a larger complex) would prove effective. This
viewpoint has been variously challenged, but both the empirical
record (see Huffaker et al. 1976b, p.56-59) and recent theoretical
analyses (Hassell 1978) strongly support the practice that the
introduction of a complex of highly host specific species, on trial,
is very unlikely to have a detrimental effect and will often give
an improved chance of success (see further Section III).

As Huffaker et al. (1976b) state,

The essential feature is to distinguish between rather
highly host-specific natural enemies and the more general
ones, and between ones which act more catastrophically
and those which act in a more stable, reliable manner.
There are, of course, two aspects of this question. One

has to do with the geographical coverage of the hosts'
environment. It is unlikely that a single enemy species
would be the superior one over an extensive range. The
value to be gained from a complex of enemies, such that
better control would be obtained over wide geographical
areas (with different enemies superior in different
situations), is largely ignored in criticisms leveled
against current practices.

The empirical record itself furnishes the clearest
basis for an evaluation of this question. This record
overwhelmingly supports the practice of introducing a
complex of primary highly specific parasites and predators. Moreover, there has been no general indication
of detrimental consequences that might ensue from introducing the more euryphagous species. Yet this record
is not entirely free of warnings relative to less host-
specific parasites and predators, and especially pathogens. The experience with pathogens has been much less
extensive, and by their very nature they are more "cata-
strophic" and erratic in action. (See further Huffaker
et al. 1971.)

The theoretically derived evidence on this question deals
with hypothetical, host-specific forms, and it is discussed in
the following section.

III. THE THEORETICAL BASIS OF BIOLOGICAL CONTROL

The theory of biological control, an economic, practically-
centered concept, is much more than a consideration of theoretical
host-parasite or predator-prey interactions, as if these interactions
occur without significant intrusions from factors external to the
interaction itself (with influence only by the constraints implied
by the mathematical values assigned to significant interaction
parameters). Yet theoretical considerations dealing with the
scope of possibilities inherent to the interaction can furnish
much insight into the problem. We will first discuss the theory
of biological control in more general terms; then turn to what we
may presume to be meaningful from mathematical modelling.

Huffaker et al. (1976a) summarized the principal attributes
of a good natural enemy as follows:

"The principal attributes of an effective natural
enemy may be categorized as follows: (1) fitness and
adaptability, (2) searching capacity, (3) power of
increase relative to the host (prey), (4) host specificity

and host preference, (5) synchronization with the
host and its habitat, (6) density-dependent performance relative to either or both the host's density
(including aggregation) and its own density (including
mutual interference), (7) detection and responsiveness
to the condition of the host, and (8) competitive
ability. ... Further on, it will be seen that these
criteria are not mutually exclusive: the notions
involved in one category often imply those of another
(DeBach and Doutt 1964, Huffaker et al. 1971).

Numerical response in the field, as well as functional
response, of a natural enemy to increase in prey density, is very
important and this itself is greatly dependent upon the primary
attribute---good searching capacity. Theoretically, the rate
of increase in female progeny/per female parent, generation to
generation, as related to host density furnishes a measure of this
numerical response. The parasitoid having the steeper numerical
response in nature would be expected to be the best one (Fig. 1).

In our general theory we have assumed that natural enemies
have evolved behavior consistent with their reducing their host's
density, as a consequence of their competition with other claimants

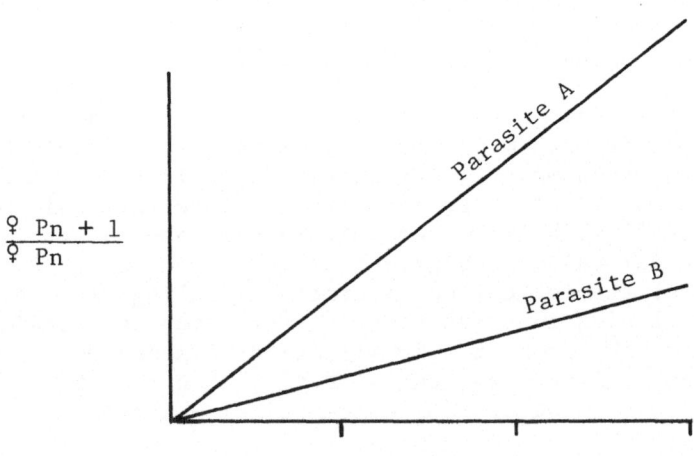

Host Density

Figure 1.   Numerical response capability of two parasite species
            (A) and (B) relative to initial host density: ♀ Pn =
            number of female parasites in the parent generation,
            ♀Pn+1 = number in the progeny generation. (From Huffaker
            et al. 1976a).

of the resource. Behavior which is overstacked in this direction
would be self-annihilative; in their behavior, there are trade-
offs and survival would seem to be paramount, whether as a sole
exploiter of the resource (host) or as a co-exploiter with some
degree of interspecific competition occurring where the resource
presents different but overlapping niches for two or more exploiters.

Bess et al. (1961) published results from establishment of a
sequence of parasitoids introduced against the Oriental fruitfly
(Dacus dorsalis Hendel) in Hawaii, which illustrates how a given
species (Opius longicaudatus Ashm.) may readily establish and be
dominant for some years. Another (O. van den boschi Full.) may be
slow to establish, but then ascend over the former dominant one
(O. longicaudatus). Finally both may be displaced by a third more
efficient species (D. oophilus Full.) introduced later.

Nextfollows an examination of recent theoretical studies
employing use of models where various attributes of (primarily)
parasitoids have been altered over a wide range in order to learn
the theoretical consequences, supposing that factors external to
the interactions modelled would not intervene. By this procedure
we can judge potential consequences, but we do not have evidence
from nature telling us the effects of other factors or that the full
range of values of the attributes used in these models is ever
realized in real populations. So we must evaluate these results
with considerable caution as to their significance in explaining
real host/parasitoid dynamics in the field.

Hassell (1978) used a two parasites/one host model to consider
the consequences of the introduction of two species as follows:

$$N_{t+1} = \lambda \, N_t f_1 \, (P_t) \, f_2 \, (Q_t)$$

$$P_{t+1} = N_t [1-f_1(P_t)] \tag{1}$$

$$Q_{t+1} = N_t \, f_1 \, (P_t) \, [1-f_2(Q_t)]$$

where N is host density, P is parasite 1 density, Q is parasite 2 density, λ is the intrinsic rate of increase of the host,
f_1 is the probability of a host not being found by P_t parasitoids
and f_2 is the probability of not being found by Q_t parasitoids.

Figure 2 presents the consequences of N, P and Q interaction, as to host equilibrium positions for various searching
efficiency values for P (i.e., a_1) and Q (i.e., a_2) used as
a ratio a_{2/a_1}.

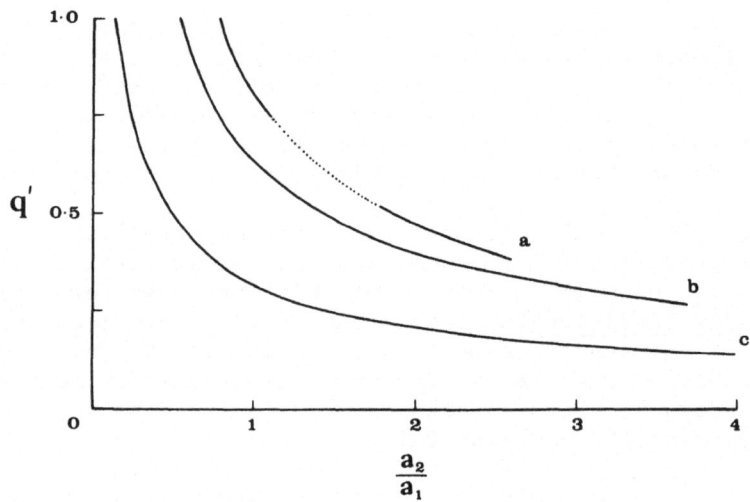

Figure 2.   Relationships from equation (1) showing the depression
            in the host equilibria (q') caused by the addition of a
            second parasitoid (species Q) in relation to the rela-
            tive searching efficiencies of $Q(=a_2)$ and $P(=a_1)$.
            (a) $k_1 = k_2 = 0.6$;   (b) $k_1 = k_2 = 0.4$;   (c) $k_1 = k_2 = 0.2$.
            (From Hassell 1978.)

        Clearly, the searching efficiency (a_2) of Q parasites in-
creases with the horizontal scale (relative to (a_1) that of P
parasites) and for each curve there is a depression of the equilibrium with increase in a_2/a_1.   The symbols k_1 for P and k_2 for
Q are exponents of their negative binomial distributions.   For
(a): $k_1 = k_2 = 0.6$; for (b), $k_1 = k_2 = 0.4$; for (c) $k_1 = k_2 = 0.2$.   The
greatest depressions in host (and parasitoid) density is with
lower values of k, and this is counteracted by the absolute values
of the host equilibrium tending to be higher with increasing k
values.   The dotted line in (a) is a region where the three species
interaction is always unstable.

        Hassell (1978) then similarly examined the possibilities for
establishing a parasitoid where a primary parasitoid was already in
equilibrium with the host (Figure 3).   The vertical axis q' is
now the ratio of host equilibria after and before the hyperparasitoid's introduction.

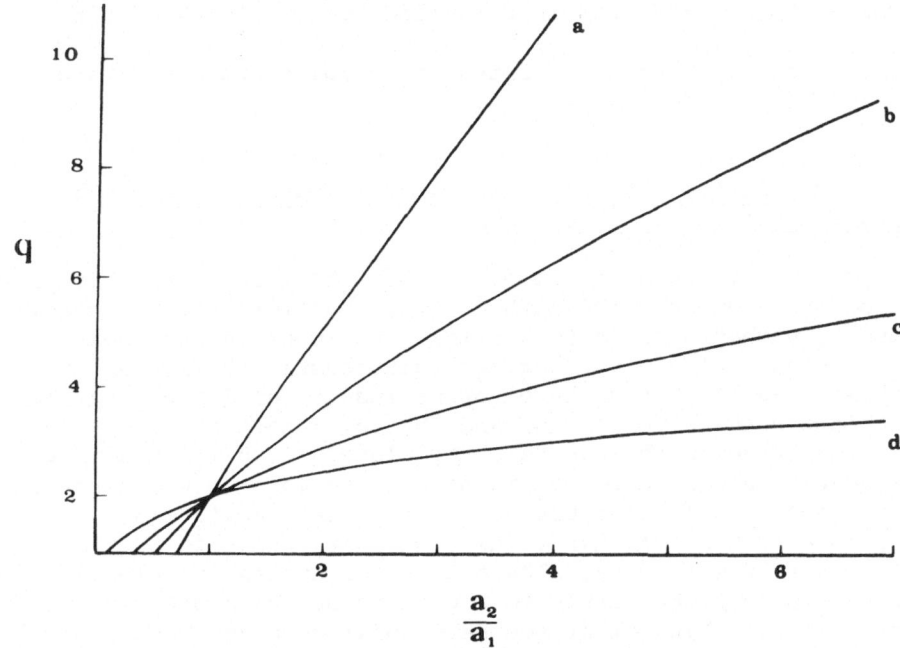

Figure 3.   As Figure 2 but not obtained from another equation where
            species Q is a hyperparasitoid.   (a) $k_1 = k_2 = 0.5$;
            (b) $k_1 = k_2 = 0.4$;   (c) $k_1 = k_2 = 0.3$;   (d) $k_1 = k_2 = 0.2$.
            (From Hassell 1978.)

        It is seen that the establishment of a hyperparasitoid always
increases the host equilibrium.   In this case   Q   is a hyperparasitoid and a_2 is its searching capacity.   Furthermore, Hassell
(1978) showed in other analyses that the possibilities favorable
to establishing a hyperparasitoid are much broader than for a
second competing primary parasitoid.

        Hassell concluded that the past practices of biological con-
trol workers of excluding hyperparasitoids but of introducing a
complex of primary parasitoids are well founded in theory, and he
pointed to parameters to look for in prospective introductions.
The work of Hassell and colleagues places us a leap further in our
confidence in making introductions, and quantitatively confirms a
number of natural enemy attributes (above) of special value: i.e.,
searching capacity, mutual interference, and host density responsive
aggregation.

IV. SOME CLASSICAL EXAMPLES ILLUSTRATING ECOLOGICAL PRINCIPLES

In this section three ecological principles are illustrated
by examples of classical biological control.

A. Plant Populations, No Less Than Insect Populations, May Be Regulated by Their Insect Predators.

Harper (1977) recently commented that the work on the biological
control of St. Johnswort (Klamath weed) in California and Australia
has been "...perhaps the most exciting experiment in the whole
science of plant-animal [population] relationships." It has, in
fact, shown clearly that an herbivorous insect can indeed regulate
a given plant species population and thereby change the whole pattern of vegetation in the habitat. By closely detailed study, the
reasons have been uncovered why in some situations (e.g., parts of
Australia) the leaf-feeding beetle, Chrysolina quadrigemina Suffr.,
does not have such a singular regulating role, while in California
and adjacent states it does. The pattern of success is closely tied
with adaptations of the beetle to a climate having a wet winter and
dry summer climate typical of southern California and West Australia.
The beetle is much less effective where summer rain occurs, both
because of high mortality, and because under even slight rains
in summer, the weed can recover quickly (Table 2).

The power of the moth, Cactoblastis cactorum Berg., to regulate
populations of Opuntia, e.g., in Queensland (Dodd 1940), formed the
backbone for rapid expansion of a worldwide effort in biological
control of weeds, and even recently, greater detail has been furnished concerning the intricate behavioral performance of this
insect (Caughley 1976).

Of more recent interest has been the demonstration that an
herbivorous insect can become so well adapted to an aquatic plant
host as to serve as an efficient regulator and control agent. Thus
the beetle, Agasicles hygrophila Selman and Vogt, and the moth,
Vogtia malloi Pastrana, introduced from Argentina to control alligator weed, Alternanthera philoxeroides (Mart.) Griseb., in south-
eastern U.S.A. are highly adapted to this emergent aquatic and
have effected practical control over a vast area. Andres (1977)
reported that the combined effects of A. hygrophila and V. malloi
have allowed a great reduction in costs of applying herbicides to
control this pest, as shown in Table 3.

B. A Complex of Natural Enemies May be More Effective Than the Best One Alone.

The complex of a parasitoid and a pathogenic virus often has
combined to produce striking, continuing biological control and

Table 2. Mortality of Wort Plants Completely Defoliated by
          Larval Stages. (From Huffaker 1967.)

| Location | Number of plants | Percentage of Plants Dead and Recovered | | |
| | | Dead | Recovery | |
			Slight	Moderate to good
CALIFORNIA				
Blocksburg 1949	244	100.0	--	0.0
Loomis 1950	15	100.0	--	0.0
Ft. Seward 1950	122	100.0	--	0.0
Blocksburg to Alder Point 1950				
Site 1	189	100.0	--	0.0
Site 2	256	100.0	--	0.0
Site 3	274	99.0	--	0.0
Site 4	256	99.0	--	1.0
Site 5	136	100.0	--	0.0
Site 6	280	100.0	--	0.0
Ft. Seward,				
Eel River 1964-65	84	94.0	6.0	0.0
1965-66	36	88.9	11.1	0.0
Cool 1964-65	13	100.0	0.0	0.0
Loomis 1964-65	29	100.0	0.0	0.0
Pilot Hill 1964-65	125	100.0	0.0	0.0
McCann 1965-66	128	75.0	25.0	0.0
				AVE.=0.1
AUSTRALIA				
Tumbarumba, N.S.W.	100	33.0	22.0	45.0
Mudgee, N.S.W. Moore-Upper	100	21.0	33.0	46.0
Mudgee, N.S.W. Moore-Lower	100	43.0	17.0	40.0
Mudgee, N.S.W. Abattoir property	100	71.0	11.0	18.0
Eurobin, Victoria	100	40.0	42.0	18.0
Bright, Victoria Smoko-dredge	100	60.0	12.0	28.0
Freeburg dredge, Victoria	100	45.0	14.0	41.0
Buckland Valley, Victoria	100	40.0	42.0	18.0
Batlow, N.S.W.	100	23.0	12.0	65.0
Tuena, N.S.W.	100	11.0	9.0	80.0
Lake Eucumbene, N.S.W.	100	0.0	15.0	85.0
Mundaroo, N.S.W.	100	34.0	30.0	36.0
Rosedale, N.S.W.	25	64.0	24.0	0.0
Warrenbayne, Victoria				
Site 1	100	7.0	24.0	69.0
Site 2	100	2.0	22.0	76.0
Site 3	100	6.0	28.0	66.0
				AVE.=45.7

balance at very low endemic status as with the European spruce
sawfly, Diprion hercyniae (Hartig), in the Maritime provinces of
Canada (Neilson and Morris 1964; Neilson et al. 1971). In this case
a baculovirus was inadvertently introduced along with parasitoids for control of this pest. The former outbreak levels were
brought down initially by an epidemic of the virus (Balch and
Bird 1944; Neilson and Morris 1964). The latter authors considered
that maintenance of the subsequent low densities of the pest was

Table 3. Some Economic Benefits from Biological Control of
          Alligator Weed. (Adapted from Andres 1977.)

Location	Changes re: Herbicide Treatments	
	Before	After
Semmes Lake, Ft. Jackson		
South Carolina	$7,500/year cost	$600/year cost
Total acreage under Army	[8,827 hectares]	[2,264 hectares]
Corps of Engineers	acres treated	acres treated
	($400,000 savings/year)	

[One-time cost of biological control introduction program = ca.
                                                    $1 million.

mostly due to the parasitoids, Drino bohemica (Mesn.) and
Exenteron vellicatus Cush. Neilson et al. (1971) considered that
either the virus or the parasites were able to control the pest
without the other, but both acting together gave a higher degree
of control. Franz (1971), citing earlier Canadian work, concluded
that the parasites, especially Exenteron because of its high
searching capacity, functions well in the spread of the virus at
sawfly densities too low to self-generate virus maintenance and
epidemics. A second well-documented example concerns the biological control of olive parlatoria scale in California, which
case is discussed in connection with the next topic.

C.   Compensating Action or Competitive Displacement May be Exhibited
When Two or More Natural Enemies Interact with a Common Host.

     Another well documented example showing that two natural
enemies (here two parasitoids) may be more effective, even in a
given locale, than the 'best' one alone, is that of control of olive
parlatoria scale, Parlatoria olea (Colvée), in California by an
ectoparasitoid, Aphytis maculicornis (Masi), and an endoparasitoid,
Coccophagoides utilis Doutt (Huffaker and Kennett 1966). Phenomenal
control was achieved by this combination even though the most
efficient species, A. maculicornis, when present alone failed to
give economically satisfactory control in some locations and years,
and C. utilis when present alone failed entirely. The pest slowly
regained devastating density (Huffaker and Kennett 1966). This
example illustrates the principles of compensation and of competitive displacement of inferior species by more efficient ones.

With host specific parasitoids, competitive displacement does
not appear to occur unless the displacer reduces the density of the
host to a level too low to allow survival of the displaced species
(the displacer is the better searcher, etc.). There have been many
examples of this and in no recorded case has such displacement led
to a reduced degree of biological control (see "Competitive displacement" in the index of Huffaker and Messenger 1976).

Even before establishment of C. utilis, several native and
introduced species that were active against olive parlatoria scale
had been virtually displaced by A. maculicornis; yet there was still
an opportunity for an efficient, well-adapted species to enter the
scene, establish readily and produce a reliable, strongly compensating role during the season of inadequacy of the other. The
two species combined achieve a more satisfactory control (at no
continuing cost) than can be accomplished by any insecticide, particularly in large, dense trees. The current density (1978) is
some 3 to 4 scales per 1000 leaves, with about half of these
parasitized.

D.   Genetic Constitution of Imported Stock is Important

·   Some recent examples illustrate the logical and long recognized
principle that a group of sibling species or ecotypes of a species
may exist over a broad heterogeneous species distribution, and that
each ecotype is especially adapted to the meterological and biological complex prevailing in its area. Introduction of a complex
of types over broad heterogeneous regions thus offers a better
opportunity for satisfying the need of all the areas infested by
the target host. Exploration efforts for biological control are
wisely concentrated in geographic areas of the home region of
the target species that are most nearly similar in both physical
and general physiognomic conditions to the target area. This
feature was stressed by Clausen (1936) but has rather recently
been clearly illustrated. Foreign explorations to locate effective
enemies of olive parlatoria scale in California resulted in
colonization efforts for a large complex of Aphytis species and
several species of other genera. Aphytis maculicornis from Spain,
Italy, Egypt, India and Persia (Iran) were tried, but only the
Persian biotype proved effective (C, above).

This example and general comparisons of the climate in Iran
to the interior valleys of California led Robert van den Bosch
to concentrate efforts to find effective parasitoids of alfalfa
weevils, walnut aphids and some other pests in that country. In
1959 Trioxys pallidus (Hal.) had been introduced into California
from France against the walnut aphid, Chromaphis juglandicola
(Kalt.). It established itself and spread rapidly in southern

California but not in the north or central interior valley. Van
den Bosch then collected this species in Iran and it has subsequently
become established in all the walnut growing areas and has accomplished complete biological control (van den Bosch et al. 1979,
R.M. Nowierski, pers. comm.).

A similar effort has been launched to find effective parasitoids of the Egyptian alfalfa weevil, Hypera brunneipennis (Boh.),
now a pest in California. This example is complicated by two
features: the specific requirement in meeting the climatic challenge
through diapause in hosts and parasitoids, and the potential in
hosts to develop resistance to parasitization by encapsulation.
For example, the types of alfalfa weevils and the biotypes of
parasites of the same species vary in their host encapsulation
potentials and in their diapause requirements. Bathyplectes
curculionis (Thom.) in Egyptian H. brunneipennis in Egypt were not
found to be encapsulated, but when tested against this species in
southern California, from 15 to 50 percent were encapsulated (van
den Bosch 1964). The Utah strain of B. curculionis which attacks
H. postica in northern California, was encapsulated to an extent
of 94 percent (Salt and van den Bosch 1967). Again van den Bosch
and Daniel Gonzalez have sought better adapted strains of several
parasites by searching in Iran, Egypt and other Near East countries.
Recently, a strain of Bathyplectes anurus (Thom.) from Iran, and
one of B. curculionis from Egypt, which are little if at all
encapsulated by local California H. brunneipennis, have been
released and the first appeared to be tentatively established in
1973. Other species are also showing promise (Hagan et al. 1976),
but as yet there has been no solid establishment. In fact, the
former promise has dimmed due to severe habitat disturbance.
Massive releases will probably be required to really test whether
any of these apparently adapted species can become established.

E.   A Parasitoid May Adapt to and be Effective Against an Alien
Host

A parasitoid may occasionally adopt a taxonomic category of
host alien to it, yet achieve a good degree of biological control.
Pimentel (1963) has, I think, overly favored this approach. It is
illustrated by the considerable control of the sugarcane moth borer,
Diatraea saccharalis (F.), in Barbados by the parasitoid, Apanteles
flavipes (Cam.), introduced from Asia. A. flavipes in Asia attacks
other genera of graminaceous borers (e.g., Chilo and Proceras)
rather than Diatraea. Apanteles diatraeae Mues. on the other hand,
is a specific Diatraea parasitoid in the greater Antilles, southern
U.S.A. and Central America. Yet A. diatraeae has failed in Barbados,
Cuba and other places to achieve biological control, while at least
in Barbados A. flavipes has succeeded (Alam, Bennett and Carl
(1971)).

The classical example of biological control of prickly pears, Opuntia inermis D.C., and O. stricta Haw. in Australia was achieved by introduction of the phycitid moth, Cactoblastis cactorum, from Argentina. This insect does not occur in southwestern North America, the native home of the invading pest Opuntias in Australia; rather it is native to and attacks related Opuntias native to Argentina, from which is was imported into Australia.

With pathogens, the control effect may be much greater on a related biotype or sibling host species than on the one where the pathogen is collected. The eradication of the dominant chestnut from the eastern deciduous forest climax by accidental introduction of the chestnut blight organism, Endothia parasitica, is a case in point. (Recent studies using hypovirulent strains of this pathogen to displace it are intriguing.)

V.   MANIPULATIONS TO CONSERVE AND AUGMENT RESIDENT NATURAL ENEMIES

There are a variety of ways by which natural enemies may be conserved or augmented. They can be conserved simply by stopping the use of pesticides inimical to them or their hosts, or by pre-serving the habitats or subsidiary natural food sources which they require. Their conservation can also be assisted by various ways of augmenting their numbers or environmental features they require. These have been discussed in detail by Rabb et al. (1976) and Ridgway and Vinson (1977); hence, will not be dealt with in detail here. Rather, a few examples of extensive practical use of these maniputions will be given.

A.   Hippodamia

The ladybird beetle, Hippodamia convergens Guer.-Men., for many years has been collected by the millions in California's Sierras and sold to growers of various crops in California, Arizona and else-where. However, it has been shown that when collected these beetles are in a state of arrested reproduction (reproductive diapause) and generally soon fly away upon release in the fields. Sometimes, enough resident, reproductive beetles can be found that the grower feels he is buying useful beetles. Now, however, certain commercial insec-taries are pre-feeding these wild collected beetles under conditions which awaken the reproductive condition (DeBach 1974), and only such beetles should be bought.

B.   Trichogramma

Mass production and release of Trichogramma for parasitization of the eggs of a wide variety of insect pests is widely practiced in the USSR, China, Mexico and some countries in South America.

Sometimes there has been success and sometimes utter failure. Much
of the work has not been adequately evaluated for effectiveness,
but enough data exist to indicate substantial effectiveness of some
programs. USSR and China workers consider that a main reason for
the failures has been that many species and strains of Trichogramma
are highly host specific to particular pest species or host habitats,
and that the wrong ones have often been used for the particular
situation (Beglyarov and Smetnick 1977).

An experimental study by Stinner et al. (1974) exemplifies
the possible benefits from mass releases of Trichogramma for control
of Heliothis spp. in Texas cotton. Table 4 details the results and
shows that very large numbers were required to achieve a higher
percentage of parasitism. Pacheco (1971) and Pacheco et al. (1971)
found that releases of up to 3 million Trichogramma/hectare in cotton
for control of bollworm in Sonora, Mexico increased parasitism up
to 84%, although with variable results and some interference from
insecticide drift. Extensive releases of Trichogramma have been
made in Mexico since 1964. By 1975, about 28 billion Trichogramma
had been released in cotton and other crops (Jiminez 1975).

In the USSR, mass releases of parasitoids and predators have
increased greatly since the early 1960's. In 1976 they were
released as a routine in more than 10 million hectares of various
crops. Many of these releases were of Trichogramma, which are used
for control of 15 pest species. Studies over many years in the
USSR, beginning in 1913, had led to knowledge of and experience
with many species and strains, at first thought to consist of a
single species, T. evanescens Westw. Trichogramma species are
reported to be effective over a vast region in the U.S.S.R. embracing
the main areas of trouble from cutworms and corn borers. Included
are the forest-steppe in the Ukraine, the foothills of the northern
Caucasus and forest-steppe areas along the Volga. Rates of release
of a corn borer race against corn borer of 30,000 to 50,000 per
hectare reduced damage by 50 to 60% (Beglyarov and Smetnik 1977).

Releases of Trichogramma in China may well exceed those in the
U.S.S.R. Huffaker (1977) reported that species of Trichogramma
were being used in some 1,700,000 acres of cotton alone in a given
year. In the important Kwangtung Province, they were being used
against leaf rollers in rice on about 1/5 of the total acreage, and
to some extent in 2/3 of the counties. In Hunan, releases were
also made against rice leaf rollers and 70 to 85% parasitization
was reported. The rates of release were determined by the density
of leaf roller eggs: if host egg densities were less than 300,000
per acre, 60,000 Trichogramma per acre were used, while 180,000
to 240,000 Trichogramma per acre were released if egg densities
were greater than 600,000 per acre. Usually such releases were
made two or more times during a season. A major advantage that

Table 4. Results of releasing <u>Trichogramma</u> on cotton for control
of bollworm and tobacco budworm.*

	Percentage of egg parasitism	
<u>Trichogramma</u>/acre*	Release	Control (pre-release values)
19,000	33	5
77,000	55	5
155,000	59	5
387,000	81	5

*"Several" releases at these rates were made

China has which makes such releases feasible is its massive manpower.

C.   Predatory Mites

Programmed releases of predatory mites for mite control as
well as other natural enemies for control of other pests, in glasshouses, has also become a widespread practice in Europe and parts
of Asiatic USSR. (See Section VI.)

D.   Behavioral Chemicals

Behavioral chemicals are also available to augment the effectiveness of predators and parasitoids. Vinson (1977) summarized
this exploratory work, which has been mainly with parasitoids,
and this brief account is taken from his review. While it is
likely that at some point search of parasitoids may involve random
search to a position where guiding stimuli are encountered, the
concept of using behavioral chemicals depends on the assumption that
their search is not entirely random. "Trails" may be reached by
random encounter, but the parasitoid may then be guided to the host
by guiding stimuli.

Kairomones are a type of chemical messenger produced by the
hosts which informs a searching parasitoid that a host is in or
has been in the area. This triggers a more intense or persistent
search in the area. Such behavior helps explain the important
density dependent aggregation of action referred to above. Application of kairomones to infested crops has been suggested as a
means of obtaining higher parasitization at the time; however,

some problems are posed with this approach (Vinson 1977) and there
has not yet been any commercial application. This work is in its
infancy, but a number of advantages to classical biological control
establishment by the employment of kairomones and other behavioral
chemicals have been suggested (see Vinson 1977).

E.   Release of The Pest Itself

Release of the pest itself in order to carry a parasitoid or
predator over a season of absence of the target host in a suitable
stage, or merely to pre-establish a parasitoid or predator population in the crop at an early time has been suggested. Huffaker
and Kennett (1956) tried this approach and demonstrated its potential
feasibility for control of cyclamen mites on strawberries but the
practice was not adopted by growers. Likewise, Parker (1971) showed
that good control of cabbage butterfly larvae in Missouri could
be obtained by field releases of small numbers of the pest butter-
flies during a critical period in summer, combined with releases of
both Trichogramma evanescens and Apanteles rubecula Marsh. Again,
this method has not been further developed and adopted by growers.

For China, however, Huffaker (1977) reported that this procedure is accepted in purple lac culture. A moth, Eublema amabilis
Moore, eats the lac and can cause serious losses. A parasite,
Microbracon greeni Ash., is released for its control. However,
again there is a period in the year when, because of lac harvesting
and seasonal progression, the parasite has too few hosts to sustain
itself, so light infestations of the moth are initiated on the
trees to bridge this gap. In one study, parasitism in pest-
augmented areas was 54% while it was only 7% in non-augmented
areas. Lac yield in parasite and pest-augmented areas increased
20 to 40% over non-release areas and the quality of lac produced
was twice as good.

The method used in integrated pest management of glasshouse
pests wherein the pest is introduced first is another example of
this technique for establishing a beneficial pest-natural enemy
interaction (see below).

VI.   PROGRAMS INTEGRATING CHEMICAL AND BIOLOGICAL CONTROL

There are many examples wherein integration of biological and
chemical control is employed; in fact biological control and pest
resistant varieties serve as the central feature of most non-Asian
programs of integrated pest management, with these integrated
closely with selective use of pesticides. In China, cultural
methods occupy this central role (National Academy of Sciences
1976).

Most cases of integrated pest management, in fact, involve
restoration of biological control disrupted by unilateral programs
of chemical pest control. Many also involve pesticide-induced
pests as central features as well as ones which have developed a
high resistance to the materials used against them.

A   Integrated Pest Management of Glasshouse Pests

Glasshouse pest management is a good example of the foregoing.
Spider mites in glasshouses in U.K. and some other European countries
had developed a high resistance to acaricides, and this led N.W.
Hussey and associates to explore the possibilities of using predatory
mites, Phytoseiulus persimilis A.-H., to control them, especially
on cucumbers. The greenhouse whitefly, another serious glasshouse
pest, had long been known to be heavily parasitized by the parasitoid
Encarsia formosa Gahan, while the aphid, Myzus percicae (Sulz.),
also had become a major pest in glasshouses, and it too can be
effectively parasitized. Table 5 from Hussey and Scopes (1977)
lists these pests, the natural enemies used against them, and the
glasshouse crops in U.K. against which they are used.

Careful attention to the biologies of the hosts, their natural
enemies, the glasshouse crop, and glasshouse temperatures gradually
have led to methods by which these pests can be largely controlled.
Usually this can be done by releases of natural enemies, with only
minimal and carefully programmed use of chemicals for other pests

Table 5.   Glasshouse Pests and Their Natural Enemies
             (From Hussey and Scopes 1977.)

Natural Enemy	Host Insect	Crops
Encarsia formosa	⊥rialeurodes vaporariorum	Tomatoes Cucumbers Sweet Peppers
Phytoseiulus persimilis	Tetranychus urticae	Tomatoes Cucumbers Sweet Peppers Chrysanthemums
Aphidius matricariae	Myzus persicae	Sweet Peppers Chrysanthemums

in the glasshouses. Information is furnished users on the expected
effects of various pesticides that might be employed in combination
with natural enemies in such operations. Yet, many problems remain,
and an intensive continuing research is required for further
acceptance of the method and to provide needed changes in procedure.
Currently, however, the method is rather commonly used in U.K.,
the Netherlands (Anon. 1975, 1976a, 1976b, Hussey and Scopes 1977),
Finland (Markkula et al. 1972) and parts of the USSR, for example.

B.   Integrated Pest Management of Pests of Oil Palms in Southwest
Asia.

       Wood (1971) has described the development of integrated con-
trol programs for various pests of oil palms, mainly in Malaysia.
Bagworms, especially, had caused much havoc in West Malaysia
beginning in the late 1950's. This seemed to be occasioned by use
of insecticides which disrupted action of their natural enemies,
perhaps aggravated by more obscure environmental factors (Bennett
et al. 1976). The overriding importance of protecting the natural
enemies has been well demonstrated, not only for these bagworms
but also for caterpillar outbreaks in Sumatra and Sabah, East
Malaysia (Wood, 1971; Wood and Nesbit, 1969; Hutauruk and Situmorang, 1971; Sankaran and Syed, 1972; Syed and Pang, 1972). This
protection, while allowing for some use of insecticides, has been
furnished by use of selective materials such as trichlorphon or
dieldrin in low volume rather than DDT, or dieldrin or endrin at
heavy dosage, for example.

C.   Integrated Pest Management of Cotton Pests in Texas and Arkansas

       The short season IPM cotton production system for South Texas
is designed to avoid the prolonged exposure of the crop to boll
weevil and·to get the crop off early so that overwintering of boll
weevils will be greatly reduced. This is combined with insecticide
treatment of weevils destined to overwinter (in diapause) so that
the weevil population is slow to develop in the following season.
Hence treatments in the summer can be delayed and natural enemies
of bollworms and tobacco budworms can be active and keep those
pests under control. Such a system has not only restored biological
control of the devastating budworm and bollworms, but the costs of
insecticide has been reduced. Yield has been increased 30%, pesticide use decreased 27%, costs per pound of lint decreased 29%, energy
use per pound of lint decreased 48%, and profit increased from
$12.40 to $104.97 per acre, a 746% increase (R.D. Lacewell, pers.
comm.)! There are similar successful programs in other regions
of Texas, but this example will have to suffice here.

In Arkansas, the main pest of cotton is bollworm, and use of careful monitoring and application of insecticides only as needed has made it possible to reduce use of conventional insecticides from 10 treatments to one in 1977 and 14 to two in 1978. One additional microbial treatment was made in 1977, while two were used in 1978. These programs have been completely adopted in one 60 sq. mile community area in 1977. An additional region composed of approximately 200 sq. miles was added in 1978 (J.R. Phillips, pers. comm.).

Table 6 presents the benefits reported for the 1977 results (1978 results are not yet available). Reduction in use of conventional organophosphate insecticides was 87% and overall savings, including costs of the microbial application, were $20 per acre. Conventional costs were $28 per acre compared to $8 for this program (J.R. Phillips, pers. comm.).

D.   Biological and Integrated Control of Sugarcane Pests in Brazil

According to K.S. Hagen, over two million hectares of sugarcane are grown in Brazil. The fields grown under both government and private company management are regularly monitored for population abundance of at least seven different insect pests. For example, in the State of Alagoas (north-east Brazil) 15 persons monitor pests in 257,000 hectares of sugarcane. Thus, one person covers about 17,000 hectares. Where need is indicated, releases of parasitoids are made against sugarcane borers (Diatraea spp.) while the fungus, Metarrhizium anisopliae, is applied against several cercopid pests. Occasionally, insecticides are also recommended for cercopids.

The release of parasitoids is required because parasitoids are disrupted during preharvest burning off of the foliage which is severe on many natural enemies in sugarcane. Furthermore, the vastness of this monoculture allows for few peripheral refuges or

Table 6.   Coy Community Benefits from IPM in Arkansas for 1977.
           (J.R. Phillips, pers. comm.)

Per Acre Figures (13,000 Acres)		
	Coy Community	Non-IPM Area	Difference
Organophosphate	0.92 lbs.	7.0 lbs.	6.1 lbs.
Costs	$8.00	$28.00	$20.00
Equivalent Fuel	1 Gallon	7 Gallons	6 Gallons
Energy	33,000 Kcal	231,000 Kcal	198,000 Kcal

alternative hosts which might have sustained natural enemies in higher numbers. Also, even occasional use of insecticides against cercopids eliminates many parasitoids of sugarcane borers. This necessitates subsequent releases of borer parasites.

Both the national producers and private companies operate insectaries in many states. The first ones were to mass culture three species of tachinid parasites of sugarcane borer; however, many new insectaries have been built since the successful establishment of Apanteles flavipes in north-east Brazil in 1974. By 1977 there were 22 insectaries in all states. In the State of Alagoas, seven insectaries were in operation and produced over 10 million A. flavipes in 1977. These were reared on sugarcane borers which in turn were cultured on artificial diets. Another million field-collected A. flavipes were also released. In addition about 50,000 tachinids were released. In 1977, the borer infestation rate was only 1.3% (% internodes bored), compared to 2.6% in 1976 and 8.5% in 1975. These rates were well below the economic threshold of 5%.

In all Brazilian states, an average infestation rate in 1975 was 5.7%; however, by 1977, it was reduced to 2.8%. There was a reduction of borer population of 47% and dead stalks were reduced by 90%.

Various cercopids are serious pests and integrated control is being developed to cope with this pest which is compatible with the biological control used against the sugarcane borer. Several laboratories mass produce Metarrhizium fungus which is sprayed on cercopid infested cane. This program is still being evaluated (K.S. Hagen, pers. comm.), and the success achieved with biological control of borers by parasitoid releases is stimulating further research on use of various natural enemies.

REFERENCES

Adkisson, P.L., 1971, Objective uses of insecticides in agriculture, in: "Agricultural Chemicals: Harmony or Discord for Food, People and the Environment," J.E. Swidt, ed., Division of Agricultural Sciences, University of California, Berkeley.

Alam, M.M., Bennett, F.D., and Carl, K.P., 1971, Biological control of Diatraea saccharalis (F.) in Barbados by Apanteles flavipes Cam. and Lixophaga diatraeae T., Entomophaga, 16: 151-158.

Andres, L.A., 1977, The economics of biological control of weeds, Aquatic Botany, 3: 111-123.

Anon., 1975, Biological pest control: Rearing parasites and predators, Growers Bull., 2, Glasshouse Crops Research Inst., Littlehampton, Sussex.

Anon., 1976a, The biological control of cucumber pests, Growers Bull., 1, Glasshouse Crops Research Inst., Littlehampton, Sussex.

Anon., 1976b, The biological control of tomato pests, Growers Bull., 3, Glasshouse Crops Research Inst., Littlehampton, Sussex.

Balch, R.E., and Bird, F.T., 1944, A disease of the European spruce sawfly, Gilpinia hercyniae (Htg.) and its place in natural control, Sci. Agric., 25: 65-80.

Beglyarov, G.A., and Smetnik, A.I., 1977, Seasonal colonization of entomophages in the U.S.S.R., in: "Biological Control by Augmentation of Natural Enemies," R.L. Ridgway and S.B. Vinson, eds., Plenum Press, New York.

Bennett, F.D., Cochereau, P., Rosen, D., and Wood, B.J., 1976, Biological control of pests of tropical fruits and nuts, in: "Theory and Practice of Biological Control," C.B. Huffaker and P.S. Messenger, eds., Academic Press, New York.

Bess, H.A., van den Bosch, R., and Haramoto, F.H., 1961, Fruit fly parasites and their activities in Hawaii, Proc. Hawaiian Entomol. Soc., 17: 367-378.

Caughley, G., 1976, Plant-herbivore systems, in: "Theoretical Ecology: Principles and Applications," R.M. May, ed., Blackwell Scientific Publ., Oxford.

Clausen, C.P., 1936, Insect parasitism and biological control, Ann. Entomol. Soc. Am., 29: 201-223.

DeBach, P., 1974, "Biological Control by Natural Enemies," Cambridge Univ. Press, New York.

DeBach, P., and Doutt, R.L., 1964, Some biological control concepts and questions, in: "Biological Control of Insect Pests and Weeds," P. DeBach, ed., Chapman and Hall, London.

Dodd, A.P., 1940, "The Biological Campaign Against Prickly Pear," Commonw. Prickly Pear Board, Brisbane, Australia.

Franz, J.M., 1971, Influence of environment and modern trends in crop management and microbial control, in: "Microbial Control of Insects and Mites," H.D. Burges and N.W. Hussey, eds., Academic Press, New York.

Hagen, K.S., Viktorov, G.A., Yasumatsu, K., and Schuster, M.F., 1976, Biological control of pests of range, forage, and grain crops, in: "Theory and Practice of Biological Control," C.B. Huffaker and P.S. Messenger, eds., Academic Press, New York.

Harper, J.L., 1977, "Population Biology of Plants," Academic Press, New York.

Hassell, M.P., 1978, "The Dynamics of Arthropod Predator-Prey Systems," Princeton Univ. Press, Princeton, New Jersey.

Huffaker, C.B., 1967, A comparison of the status of biological control of St. Johnswort in California and Australia, Mushi, 39, Suppl., 51-73.

Huffaker, C.B., 1977, Augmentation of natural enemies in the People's Republic of China, in: "Biological Control by

Augmentation of Natural Enemies," R.L. Ridgway and S.B. Vinson, eds., Plenum Press, New York.

Huffaker, C.B., and Kennett, C.E., 1956, Experimental studies on predation, 1. Predation and cyclamen mite populations on strawberries in California, Hilgardia, 26: 191-222.

Huffaker, C.B., and Kennett, C.E., 1966, Biological control of Parlatoria oleae (Colvée) through the compensatory action of two introduced parasites, Hilgardia, 37: 283-335.

Huffaker, C.B., Luck, R.F., and Messenger, P.S., 1976a, The ecological basis of biological control, Proc. XV Int. Congr. Entomol., 560-586.

Huffaker, C.B., and Messenger, P.S., eds., 1976, "Theory and Practice of Biological Control," Academic Press, New York.

Huffaker, C.B., Messenger, P.S., and DeBach, P., 1971, The natural enemy component in natural control and the theory of biological control, in: "Biological Control," C.B. Huffaker, ed., Plenum Press, New York.

Huffaker, C.B., Simmonds, F.J., and Laing, J.E., 1976b, Theoretical and empirical biological control, in: "Thoery and Practice of Biological Control," C.B. Huffaker and P.S. Messenger, eds., Academic Press, New York.

Huffaker, C.B., van den Vrie, M., and McMurtry, J.A., 1970, Ecology of tetranychid mites and their natural enemies: A review, II. Tetranychid populations and their possible control by predators: An evaluation, Hilgardia, 40: 391-458.

Hussey, N.W., and Scopes, N.E.A., 1977, The introduction of natural enemies for pest control in glasshouses: Ecological considerations, in: "Biological Control by Augmentation of Natural Enemies", R.L. Ridgway and S.B. Vinson, eds., Plenum Press, New York.

Hutauruk, C., and Situmorang, H.S., 1971, Some notes on the control of the bagworm, Metisa plana Wlk., in North Sumatra, in: "Crop Protection in Malaysia," R.L. Wastie and B.J. Woods, eds., Inc. Soc. of Planters, Kuala Lumpur.

Jiminez, J., Eleazar, 1975, The action program for biological control in Mexico. Presented at a work conf. of the U.S. and Mexico, "Biological control of pests and weeds," Brownsville, Tex. (Feb. 20-21, 1975).

Laing, J.E., and Hamai, J., 1976, Biological control of insect pests and weeds by imported parasites, predators, and pathogens, in: "Theory and Practice of Biological Control," C.B. Huffaker and P.S. Messenger, eds., Academic Press, New York.

Markkula, M., Tiittanen, K., and Nieminen, M., 1972, Experiences of cucumber growers on control of the two-spotted spider mite, Tetranychus telarius (L.), with the phytoseiid mite, Phytoseiulus persimilis A.-H., Ann. Agric. Fenn., 11: 74, 78.

McMurtry, J.A., Huffaker, C.B., and van den Vrie, M., 1970, Ecology of tetranychid mites and their natural enemies: A review. I. Tetranychid enemies: Thier biological characters and the impact of spray practices, Hilgardia, 40: 331-390.

National Academy of Sciences, 1976, "Insect Control in the People's
    Republic of China," CSCPRC Rep. No. 2, National Academy of
    Sciences, Washington, D.C.
Neilson, M.M., Martineau, R., and Rose, A.H., 1971, Diprion hercyniae
    (Hartig), European spruce sawfly (Hymenoptera: Diprionidae).
    Commonw. Inst. Biol. Control Tech. Commun., 4: 136-143.
Neilson, M.M., and Morris, R.F., 1964, The regulation of European
    spruce sawfly numbers in the Maritime Provinces of Canada from
    1937 to 1963. Can. Entomol., 96: 773-784.
Pacheco, M., Francisco, 1971, Evaluation of the parasitism of the
    wasp, Trichogramma spp. on eggs of the bollworm in Sonora-1971,
    Prog. Rep. No. 3, National Institute for Agricultural Research
    (INIA).
Pacheco M., Francisco, Carrillo S., J.L., Monge C., J., and
    Covarrubias G., R., 1971, Adelantos sobre la evaluacion del
    parasitismo de la avispita Trichogramma spp. sobre huevecillos
    de gusano bellotero en sonora durante 1969, Agricultura Technia
    en Mexico, Organo del INIA, SAG, 3(2): 53-57.
Parker, F.D., 1971, Management of pest populations by manipulating
    densities of both hosts and parasites through periodic releases,
    in: "Biological Control," C.B. Huffaker, ed., Plenum Press,
    New York.
Pimentel, D., 1963, Introducing parasites and predators to control
    native pests, Can. Entomol., 95: 785-792.
Porter, B.A., 1947, Orchard insecticides, in: "Yearbook of Agri-
    culture", U.S. Dept. Agric., Washington, D.C.
Rabb, R.L., Stinner, R.E., and van den Bosch, R., 1976, Conservation
    and augmentation of natural enemies, in: "Theory and Practice
    of Biological Control," C.B. Huffaker and P.S. Messenger,
    eds., Academic Press, New York.
Ridgway, R.L., and Vinson, S.B., eds., 1977, "Biological Control by
    Augmentation of Natural Enemies," Plenum Press, New York.
Ripper, W.E., 1956, Effects of pesticides on balance of arthropod
    populations, Annu. Rev. Entomol., 1: 403-438.
Salt, G., and van den Bosch, R., 1967, The defense reactions of
    three species of Hypera (Coleoptera, Curculionidae) to an
    ichneumon wasp, J. Invertebr. Pathol., 9: 164-177.
Sankaran, T., and Syed, R.A., 1972, The natural enemies of bagworms
    on oil palms in Sabah, East Malaysia, Pac. Insects, 14: 57-71.
Smith, Ray F., and Reynolds, H.T., 1977. Some economic implications
    of pesticide overuse in cotton in: "New Frontiers in Pest
    Management," California State Senate Office of Research,
    Sacramento.
Stinner, R.E., Ridgway, R.L., Coppedge, J.R., Morrison, R.K., and
    Dickerson, W.A., Jr., 1974, Parasitism of Heliothis eggs after
    field releases of Trichogramma pretiosum in cotton. Environ.
    Entomol., 3: 497-500.
Syed, R.A., and Pang, T.C., 1972, Status, history and control of
    Setora nitens (Wlk.) in Sabah, in: "Cocoa and Coconuts in

Malaysia," R.L. Wastie and D.A. Earp, eds., Inc. Soc. of
  Planters, Kuala Lumpur.
van den Bosch, R., 1964, Encapsulation of the eggs of Bathyplectes
  cueculionis (Thomson) (Hymenoptera: Ichneumonidae) in larvae
  of Hypera brunneipennis (Boheman) and Hypera postica (Glyllenhal)
  (Coleoptera, Curculionidae), J. Insect Pathol., 6: 343-367.
van den Bosch, R., Hom, R., Matteson, P., Frazer, B.D., Messenger,
  P.S., and Davis, C.S., 1979, Biological control of the walnut
  aphid in California: Impact of the parasite, Trioxys pallidus,
  Hilgardia, 47: 1-13.
Vinson, S.B., 1977, Behavioral chemicals in the augmentation of
  natural enemies, in: "Biological Control by Augmentation of
  Natural Enemies," R.L. Ridgway and S.B. Vinson, eds., Plenum
  Press, New York.
Wood, B.J., and Nesbit, D.P., 1969, Caterpillar outbreaks on oil
  palms in Eastern Sabah, The Planter, 45: 285-299.
Wood, B.J., 1971, Development of integrated control programs for
  pests of tropical perennial crops in Malaysia, in: "Biological
  Control," C.B. Huffaker, ed., Plenum Press, New York.

TOWARD MORE RATIONAL POLICY

D. Woods Thomas

Executive Director
Board for International Food and Agricultural Development
Washington, D.C.

INTRODUCTION

This Boyce Thompson Dedication Symposium has focused on the issues involved in linking basic research to crop improvement programs in the less-developed countries. A number of well-known agricultural scientists have addressed several components of this question. The Institute and all those who have participated in the Symposium are to be congratulated for their efforts in this respect. I say this because I have become convinced over the past several years that the degree to which the world will be able to effectively cope with its massive food, nutrition and poverty problems will depend upon the degree to which the United States and other developed nations can come to grips with the generic policy issue involved. This is the issue of whether or not the developed nations are willing to devote, for some time to come, a significant fraction of their massive scientific capacity in the agricultural and related sciences to the systematic study of factors constraining increased resource productivity in the less developed nations.

The basis for this position is straight forward. The pervasive and fundamental problem in the developing nations is one of poverty in its purest form. There is only one viable means of increasing the levels of real income of these people in a sustained manner. This is through increasing the productivity of their land, labor and capital resources. Expanding food production by expanding the use of traditional inputs will not result in increased incomes. It turns out that, with some exceptions, most LDC resources having potential for increased productivity are in the

agricultural sector where the large majority of the world's poor eke
out their meager livelihood. Thus, the rural sectors of these
societies must be the principal point of concentration of any
rational economic development strategy.

The problem is one of identifying the particular constraints
which are fundamental to increased resource productivity, as
contrasted to increased food and fiber output. There are those who
argue that internal economic disincentatives explain some part of
the failure of LDC's to achieve greater increases in resource productivity. There are others who argue that social and cultural factors inhibit agricultural growth. Still others point to imperfections
in indigenous public and private institutions as a primary constraint
to increased productivity. There are those who argue that the
principal constraint is technical in nature; i.e., that resource productivity is bounded by technical production functions which are of
such character that output per unit of resource input would not
increase even if other constraints were removed.

All of the above have conceptual and empirical validity. The
particular significance associated with any one constraint or set
of constraints varies from country to country, region to region and
commodity to commodity. It is also quite clear that systematic
research has something of value to contribute in each of the areas.
This paper, however, will focus on technical constraints.

TECHNICAL CONSTRAINTS TO INCREASED RESOURCE PRODUCTIVITY

It is true that the upper limits of output which might be
obtained from resources used in the farm production process are given
by the known set of technical production possibilities. This is true
in all situations. It is true on subsistance farms and on commercial
farms -- on small farms and on large farms. It is true in all countries -- developed and less-developed, alike.

There is considerable empirical evidence that the technical
production functions faced by farm decision makers in the poor
nations tend to preclude the possibility of obtaining meaningful
increases in resource productivity. There is increasing evidence
that, within such technical production constraints, farmers in these
situations tend to allocate their scarce resources in an economically
rational way. The problem is that the technical constraints make
it virtually impossible for them to obtain significant increases in
output from the limited resources at their command. Increased output
is attainable only through employment of additional resources --
more land, more labor, more capital -- if such are available. This
does little toward resolving the basic poverty problem since under
these circumstances economic returns to resource inputs tend to
remain constant -- people remain poor.

It is in quantum increases in technical production possibilities that rests the potential for increasing the productivity of resources used in farm production. Upward shifts in technical production functions through technological innovation provide the basis for increased returns to farm inputs -- and the possibility of associated increases in income levels of rural people. It also provides the basis for meaningful shifts in the internal food supply functions of LDC's. This translates to reduced food costs for all consumers and, therefore, to increased levels of real income.

There are other equally important effects of technological innovation in agriculture -- the formation of capital for both farm and non-farm investment, the transfer of redundant resources out of agriculture, the expansion of demand for products of a nation's industrial sector, increased employment opportunities in the farm service sector, reliable internal sources of raw materials for industrial uses, increased foreign exchange earnings through the export of farm products, and the like. In short, technological innovation in predominantly rural, less-developed economies appears to be a necessary if not sufficient condition to more general economic development and sustained growth in these societies as well as for meeting basic food requirements.

THE ROLE OF RESEARCH

In this context, it is necessary to be explicit about a point that is often neglected or misunderstood. This is that technology responsible for making upward shifts in farm-level production does not occur by accident. It is generated in only one way, through sustained investment in research into the physical, biological, economic and social factors constraining the relevant production functions. Without such investment across the basic/applied spectrum of research, the knowledge base essential to the development of technology for increasing productivity simply will not exist. Further, given the atomistic manner in which farms are organized, worldwide, such knowledge will not be generated within the sector. Also, the benefits from new agricultural technologies usually cannot be captured by individual investors -- for the most part, such technologies are public goods. Thus, if research investment does occur, it will be via conscious public decision and through public institutions. The fundamental difference between high and low resource productivity agricultural economies is that such public decisions and subsequent investments have been made in the former, but not in the latter.

Most informed observers would agree that few, if any, of the LDC's have sufficient indigenous agricultural research capacity to generate the knowledge and technology essential to the establishment

of farm production possibilities of the order of magnitude required
to have significant impact on the internal food, nutrition, poverty
and development problems which they face. Indeed, few LDC's have
national research systems capable of utilizing effectively the
basic knowledge already available in developing countries. Difficulties also exist in the adaptation of existing production technology in a manner consistent with LDC's physical, biological, economic
and human environments. And such human and institutional research
capacity cannot be created instantaneously. It usually takes some
ten years beyond the baccalaureate degree for an agricultural
scientist capable of making significant contributions to develop.
Experience has indicated that, under the best of circumstances,
creating an effective national complex of agricultural research
institutions is a long-term endeavor requiring fifteen to twenty or
more years of concerted effort. The time and resource requirements
to do so in fifty or more LDC's around the world will be massive.

The scene that I have sketched out might cause some to throw
up their hands in dismay. However, there is an option which, if
exercised, has the potential of making the fate of the impoverished,
hungry and the malnourished around the world less dismal than it
might otherwise be. This is to bring the scientific capacity of
the agricultural research institutions in the United States and
other developed nations to bear on the technical, economic and
social constraints to increased resource productivity in the LDC's.
This will require the creation of an effective, collaborative, international research network including developed country institutions,
LDC institutions and the International Agricultural Research Centers.

During the next 20 to 30 years, the outcome of the developing
world's struggle for development and growth will be decided. In
this same period, the agricultural research community of the more
developed nations will have, hopefully, the opportunity to influence
the outcome in a critically positive way. Potential contributions
of this community are many and varied.

First, the LDC's must find a means of creating indigenous
agricultural research systems capable of meeting the never-ending
demands for new knowledge, adapted technological innovations and
related technical, economic and social information. Established
agricultural research institutions in the U.S. and elsewhere,
through appropriate collaborative work, can contribute to the
acceleration of the rate at which LDC national research institutions
and systems evolve, mature and become positive contributors, not
only to problem solution within their own national boundaries, but
also on the world scene. This type of activity, I believe, is well
within the doctrine of most such institutions. It is simply another
way of rolling back the frontiers of knowledge with a potentially
great multiplier coefficient.

Second, the creation of adequate agricultural research capacity in the LDC's will require a sharp expansion in the number of agricultural scientists to staff the institutions and conduct the requisite research programs. The agricultural research institutions of the developed nations and the international components of the worldwide network of scientific endeavor face a major challenge in training significant numbers of LDC scientists. Innovative means of developing a greatly expanded cadre of mature agricultural scientists having not only the scientific knowledge and methodological skills, but also the will, desire and philosophy to contribute to the solution of problems, must be found.

Third, this same set of institutions has the opportunity of making major direct and indirect contributions to improved agricultural production possibilities in the LDC's. These opportunities exist over the entire range of scientific endeavors in which they are engaged. The discovery of new knowledge relevant to biological, physical and economic productivity constraints has universal applicability. A number of such problem areas in which such is needed have been discussed earlier in this symposium. There are many others which would have significance to agriculture in the developing countries as well as the developed nations. In fact, it is likely that scientific breakthroughs in this realm would have greater impact, in terms of relative productivity increases, in the LDC's than in the higher resource productivity areas. It is also possible that some productivity constraints in the LDC's may yield only to more sophisticated approaches.

At other points on the spectrum, less sophisticated approaches to problems have the potential of contributing directly to the development of technologies capable of removing productivity constraints. There appears to be a fairly wide range of problem areas in which collaboration among LDC and developed country scientists would be highly complementatry in the sense of being mutually and directly beneficial.

It has been my observation that scientists working on "basic" problems are reasonably flexible vis-a-vis the particular aspect of a problem upon which they work as long as it is within the general range of their scientific interest and competence. Other things being equal, they would appear to be quite willing to apply their talents to issues having obvious potential for resolving productivity problems in the LDC's vis-a-vis those in which the connection is less clear.

It has also been my observation that many agricultural scientists interested in more applied issues are not only willing but also quite anxious to collaborate with colleagues abroad on research into problems of mutual interest. They understand the potential

complementaries of such interaction. They do less of it than they
would desire for lack of information and contacts or the existance
of institutional or fiscal constraints over which they have no
control. Improved information and communications relative to the
nature of production constraints in the LDC's coupled with the
relaxation of institutional and fiscal barriers would result in a
material increase in useful collaboration.

THE DILEMMA

     To those who count themselves among the "believers" in the
necessity of appropriate levels of sustained investment in agricultural research, it is difficult to understand why the LDC's have
not followed this pattern. It is equally difficult to understand
why development assistance "donors" did not long ago adopt similar
policies. It is also difficult to completely understand why agricultural research institutions in the United States and other
developed nations have not, in their own self interest, opted for   °
greater international collaboration on the part of their professional staffs. The logic of the arguments, the expanding body of
empirical evidence of high rates of return to such investment and
the general knowledge of the agricultural development experiences
of the more developed nations all argue strongly for such policies.

     Yet, this has not been the case. The LDC's have not counted
the creation of viable national agricultural research systems among
their top national priorities for investment of scarce public funds.
The major development assistance donors, with some exceptions, have
not held direct investments in research in very high esteem -- in
fact, there are examples of implicit if not explicit policies against
such activities. Our own agricultural research institutions have
performed little better in this respect -- doctrine, policy and
fiscal limitations have served as effective barriers to international
research collaboration on any significant scale.

     In recent years, there have been some encouraging signs. The
evolution of the complex of International Agricultural Research
Centers (IARC's) over the last 15 or so years, with steadily growing,
multi-donor support, attests to recognition of the value of research
in agricultural development. AID, foundation and other sponsorship
of university involvement in "institution building," including
research components, abroad during the 1950's and 1960's was significant. AID has maintained, for some years, a small, centrally-
funded contract and grant research program in key agricultural
problem areas. A number of the LDC's are now paying more attention
to expanding their agricultural research capacities.

     The enactment of the Title XII Amendment to the U.S. Foreign
Assistance Act provided, among other things, authorization for the

development of collaborative research among U.S., LDC and IARC agricultural scientists in problem areas of mutual interest. It also
urged greater participation of U.S. institutions in the international
network of agricultural research. The Administration's proposal,
currently before the Congress, to create the Institute for Scientific
and Technological Cooperation (ISTC) is another positive sign.

Despite the above, all efforts to bring existing scientific
capacity to bear on agricultural development problems in the LDC's,
in my judgment, have fallen far short of that which is and will be
needed. After thirty years of international cooperation to accelerate agricultural growth in the LDC's, few such countries have
anywhere near adequate agricultural research capacity. It has taken
some 15 years to obtain reasonable levels of support for the IARC's
and this is a somewhat tenuous, year-to-year, multi-donor operation.
AID's centrally-funded contract and grant research program, including
support for the new Title XII Collaborative Research Support Program
(CRSP), is miniscule; over the last five years it has accounted for
about 2-1/2 percent of the Agency's total food, nutrition and agricultural development investment. In 1979, the Agency allocated
about $13.4 million to this program and has requested $17.4 million
in 1980. For the first time in its long history, the USDA recently
obtained Congressional authorization to engage in cooperative
research with foreign agricultural scientists in cases where the
benefits of the cooperative work will not necessarily accrue to
domestic agriculture. Most seriously, despite the keen interest and
obvious need, the massive agricultural research talent housed in our
American system of higher education has not been engaged in any
significant way in the efforts of the developing countries to reduce
hunger, improve nutrition and increase income and the living stan-
dards of their people.

WHY THE DILEMMA?

All of those who have worked in the vineyard of international
agricultural research are fully aware of the difficulties involved
in obtaining resources to support this work. The truth of the
matter seems to be that public decision-makers -- in our Congress
and in its counterparts abroad, in the government bureaucracies,
in state legislatures and U.S. research institutions -- do not
believe that research has very much to contribute to agricultural
development. We need to understand why this is so.

This is an extremely complex and many-faceted question to
which there is no simple, easy answer. Let me, however, set forth
some hypotheses relative to the reasons for the lack of broad-based
support for and favorable public policies toward agricultural
research in the international context.

It is important to remember that internationally-oriented
agricultural research is a special case of agricultural research and
that all of this is but one component of science and technology,
generally. In the more developed countries, the last decade appears
to have witnessed a pervasive disaffection with scientific and
technological endeavors. The public, and the people who represent
it in public decision-making roles, have been re-examining their
value systems relative to what matters and what doesn't. In this
process, questions have been raised relative to the value that these
societies place on at least some of the products of investments in
research. I suspect that such general assessments have "rubbed-
off" and eroded somewhat the support base for agricultural research.

Several generations of ordinary citizens have lived through
extended periods of agricultural surpluses. It is difficult to
explain to those people the significance of an agricultural sector
approaching the limits of productivity and the need for greater
research investment in order to elevate such ceilings. This, too,
has taken its toll in terms of support for additional research investment.

On the international scene, there never has been a major grounds-
well of public support for foreign assistance of any kind -- let
alone research. The relationship between the development of poorer
nations and the continued well-being of the economically more advanced
nations has not been well documented or articulated. To many people,
the relationship appears nebulous, at best. This, in addition to a
tendency toward neoisolationism in some of the developed nations has
been detrimental to firm public policies reflecting support for
increased investment in internationally-oriented agricultural research.

In a more general way, I believe that those of us in the
scientific community seriously underestimate the complexity of the
research/knowledge/technology/productivity/well-being relationship
as viewed by non-scientists. To the "initiated" the linkages are
clear-cut; to the "uninitiated", they are not. Yet public policy-
makers, administrators and their staffs are, for the most part, non-
scientists; hence, the connections are not automatic. Perhaps it
should not be surprising that some of our pleas for greater research
support go unheeded!

The lack of greater support for agricultural research within
some international development donor organizations is somewhat more
complex and difficult to understand. Such organizations are staffed
by professional development personnel who should understand the
role of science and technology in the development process; yet
development assistance programs do not seem to reflect such understanding.

The answer to the anomaly probably rests in a variety of factors. On the one hand, most such agencies and organizations must
be responsive to their executive and legislative bodies. Lack of
support for research as an effective, powerful development "tool"
within these bodies tends to be reflected in the makeup of the
organizations' programs. Such organizations also must be responsive
to the agricultural development priorities and strategies of the
host nations with which they work. The leadership in most such
nations is under extreme pressure to achieve agricultural development at highly accelerated rates and in visible forms. In these
situations, it is quite understandable that they tend to seek
"short cuts" and that research may not stand very high on their
list of priorities for outside development assistance. Again, this
will be reflected in the programs of work of the development
assistance organizations.

Finally, the system of rewards and penalties for the staffs
of such agencies appear to be heavily weighted in favor of short-
run action programs and against longer-run, higher-payoff investments
which tend to characterize most research enterprises. This, too,
militates against research and tends to relegate it to being the
recipient of residual funds after other activities have been fully
funded. This happens even though the attendant resource allocation
pattern may bear little relationship to the relative productivities
of the alternative investments.

THE BOTTOM LINE

In this paper, I have argued that technical change in agriculture may well be the unique key to the eventual solution of the
hunger, malnutrition and poverty problems in the LDC's. I have
indicated a concern that insufficient resources are being devoted
to the research needed to produce the resource productivity
increasing technology required. I have lamented the fact that it
has not been possible to engage effectively the U.S. agricultural
research community in this process. I have also suggested some
possible reasons for the apparent lack of support for expanded
international agricultural research endeavors.

The bottom line is the question of what might be done to
ameliorate the somewhat dismal situation which appears to exist.
There is but one answer to this question. This is for the agricultural research community to carry its message to public policy
decision-makers and others in a far more cogent, articulate, con-
vincing way than it has in the past. Public policy makers are
reasonable people trying to do their work as effectively and cor-
rectly as they possibly can. Like all decision-makers, their
decisions and actions can be no better than the information they

have relative to the probable outcomes of alternative courses of
action. I suspect that the information they have or can obtain
relative to the value of research in the development process is both
limited and highly imperfect.

The story to be communicated is impressive. The empirical
evidence of the rates of return to investments in agricultural
research is difficult to ignore. Examples of the impact of research
on agricultural development, in developing as well as developed
countries, abound. Yet, few policy-makers have this information
in their decision-making matrices. Our community has tended to
rely too much on sporadic exhortations for the support of science
for science's sake and too little on the hard evidence of the true
value of research investments to a society. In short, I am sug-
gesting that we must find some effective, organized, sustained
means of providing factual information and analyses on this subject
to those who make the decisions with which we are primarily concerned.
If this is done well, I am convinced that the response will be
positive.

THE ROLE OF SCIENTIFIC RESEARCH IN RURAL DEVELOPMENT

Milton J. Esman
John S. Knight Professor of International Studies
Cornell University
Ithaca, New York 14853

It is an intimidating experience for a layman who is not a biologist or even a specialist in agriculture to be addressing agricultural scientists. You represent a very small scientific elite with enormous prestige, a prestige that has been richly earned, for your efforts have averted famine in large areas of the world. Your expertise, however, commits you to continuing responsibilities, for humankind is apparently condemned to an unremitting preoccupation with feeding its growing numbers, at least for the next half century.

Though recent technological breakthroughs in agricultural science have indeed been dazzling, what we know of the history of sciences, including the agricultural sciences, must caution us against high expectations for similar breakthroughs in the near future. This caution is reinforced by the judgement of most practicing scientists. Progress in science has always depended on long-term investments in scientific infrastructure and on patient, gradual, incremental additions to knowledge and to practice.

Moreover, science usually responds to demands for information from the society that supports it. This determines how problems are defined and what priorities are selected for investigation. Thus there are serious limitations to what U.S. and other western scientists in their laboratories and experiment stations and even in the international agricultural research centers can contribute to agricultural and rural development in less developed countries. They can be effective only if they can work with and through indigenous research and development institutions which are addressing needs and priorities identified by local experience,

are answerable to local publics, and are rewarded for solutions
to local problems. Scientific and technological progress in any
country depends on a domestic scientific infrastructure supported
by that society and oriented to its needs. Only indigenous
institutions can effectively translate general scientific knowledge
into information and practices suited to specific local conditions
and to particular micro-environments, especially if such knowledge
is to be adapted to the needs of small and marginal farmers.

Therefore, the most useful and long-lasting contribution that
westerners can make to scientific development in the poorer
countries is not in finding solutions to particular problems --
though this may be helpful in the short run -- but in continuing
to strengthen the research and development institutions and networks in these countries. Since scientific autarchy is a contra-
diction in terms, the research and development institutions in
developing countries must be linked to external centers of knowledge
generation, including national centers such as Cornell University
and the Boyce Thompson Institute, and the international agricultural research centers that have been established during the past
decade under western financial and intellectual auspices and staffed
primarily by western scientists. The main strategy, however, must
be the development of vigorous national institutions; my concern
is that this emphasis which was prominent in the 1960s has been
attenuated and diverted by enthusiasm for the new international
centers and the expectations that they would produce miraculous
technological breakthroughs.

Many problems remain to be worked out in the division of
labor between the international centers and the national research
and development systems, and these relationships must be dynamic,
depending on the stage of development of the national institutions.[1]
Thus, relationships between international centers and national
research systems in India and Brazil on the one hand will differ
from their relationships with Nepal and Niger. The governing principal must continue to be the gradual displacement of scientific and
technological dependency with increasing self reliance. Thus, the

_____

[1]The Consultative Group in International Agricultural Research,
(CGIAR) which finances and supports the international centers,
organized in 1977 a task force on "international assistance for
strengthening national agricultural research." Its Report in August
1978 recommended that a service for supporting national agricultural
activities be established under CGIAR auspices, because of the urgent
need to strengthen national research institutions especially on the
management side. This issue is discussed in the 1978 Report of the
International Agricultural Development Service, which is already
active in this field.

primary strategy in resource allocation for agricultural and rural development and the test of achievement among those concerned with international development must be success in building national institutional capabilities.

By rural development we refer to multi-disciplinary measures to increase the productivity and the well being of rural people, especially of the poorer and weaker strata among them. Though conditions differ widely both among and within countries, research that has been underway at Cornell indicates that the majority of rural households in most developing countries, on all continents, are headed by landless workers, tenants and small or marginal cultivators.[2] The land resources in the hands of members of this poor majority are usually meagre and of poor quality. The public services available to them are minimal and they have limited access to commercial inputs. Though all members of the household must work, including women and even children, productivity is low, labor is frequently underemployed despite seasonal migration, chronic indebtedness is common, family incomes hover around the poverty line, and large numbers live in misery, insecurity and severe deprivation -- what the World Bank identifies as "absolute poverty". Because of rapid population growth, the size of the rural labor force continues to grow, often on a fixed land base, despite large scale migration to urban centers. In India, for example, population is expected to increase from the current estimated figure of 650 million to 1 billion by the turn of the century. About half of this growing population, about 175 million, will continue to live and work in rural areas. This means that jobs for an additional 4 million workers must be found every year in the already over crowded and impoverished rural areas of India. This implies the urgent need for additional employment opportunities through the intensification of agricultural production and through the expansion of non-agricultural activities, including small industry, construction and the services.

Though the majority of people in the rural areas of developing countries live at or below the poverty line, most of the substantial benefits of crop and animal research to date have accrued to farmers on better, often irrigated, soils and with access to commercial inputs. Only a small number of major food crops have been affected. A substantial number of cultivators have benefitted significantly and this minority of "progressive farmers" have contributed the bulk of increases in agricultural production, but a much larger number of cultivators have not benefitted at all. The benefits have flowed in Latin America primarily to larger commercial

---

[2]Milton J. Esman, Landlessness and Near Landlessness in Developing Countries, Ithaca, N.Y., Rural Development Committee, Center for International Studies, 1978.

farmers, in Asia to farmers on irrigated soils, while Africa has
been largely untouched by the Green Revolution. The majority of
farmers and laborers on poorer soils and on small holdings with
limited access to commercial inputs have been unaffected by modern
agricultural research, and their numbers are increasing rapidly.

Let us assume a very favorable scenario for the next decade,
that is: a) if population growth rates should decline rapidly
(they have begun to fall in some countries, but only slowly and a
clear trend has not yet been established); b) if industrial employment should increase rapidly (there is little evidence of this in
most developing countries because of the capital intensive nature
of most industrial processes and the moderate rates of capital
formation and of industrial growth); c) if governments should invest
more heavily in agriculture and in rural development, especially
in irrigation thus correcting the pronounced urban bias in public
investment and expenditures (there is some evidence that this has
begun to happen); and d) if there should be more equitable land
tenure arrangements (here the picture is very uneven). Even under
this most favorable scenario, large numbers of cultivators, many
more than at present, will be working very small holdings on low
quality soils with limited public services and with access to minimum
commercial inputs; and increasing numbers will be landless laborers.
Unlike the experience of the U.S. and other industrialized countries,
it will not be possible to transfer large numbers of marginal or
"surplus" rural workers out of agriculture and into industrial
and urban occupation, because industrial jobs are not being created
in sufficient numbers to absorb the rapid increase in the rural
labor force. They will have to be supported on the land and in
related occupations in rural areas. Their needs will make increasing
claims on the limited resources of governments and on the sympathies
and energies of development assistance agencies. Rural development
must be the key to any humane strategy of development in low income
countries, and the emphasis must be on increasing productivity,
employment, and welfare among the poor majority.

If agricultural research is to respond to the needs of this
poor majority of small and marginal cultivators and landless
workers, it must be oriented to the policy goals of increasing
outputs and labor utilization per unit of land and to technologies
that call for limited use of expensive commercial inputs. Intensification, of course, can create serious problems, especially on
weak and low-quality soils, unless rigorous soil management methods
are used which are appropriate to the local situation. Small
holdings, however, tend to be more productive and to use more labor
per unit of land than larger holdings with equivalent agronomic
properties. Serious intensification strategies will require much
more intensive research support than has been available in the past.
It will have to be oriented in many areas to more productive <u>farming</u>

systems - to the mixture of crops and animals that a small farmer
can combine over the entire annual agricultural cycle, rather than
improved varieties of a single crop. Because there has been so
little research on farming systems in recent years, the prospect of
rapid breakthroughs is not encouraging.

How can agricultural researchers become better informed and
more responsive to the needs, priorities, limitations and possibilities of the heterogenous groups of small and marginal cultivators working in their distinctive micro-environment? The
literature reports almost routinely cases of farmers rejecting the
advice of extentionists who carry messages from agricultural
research centers. While the tendency of agricultural scientists
and administrators is to blame the farmers for conservatism and
resistance to change, closer investigation usually discloses that
their advice is inappropriate to the circumstances of the farmers
concerned. It prescribes commercial inputs that the farmers cannot
afford, it imposes risks that they are not prepared to accept,
it overlooks and dismisses information which farmers have accumulated
over years of experience, or it fails to take account of their
farming systems and their annual returns from their land and labor,
rather than from a single crop. If agricultural research is to
serve the rural majority, if it is to define the problems it works
on in terms of the concerns of small and marginal cultivators, its
point of departure must be real life farming conditions, not the
conditions of the laboratory or experiment station, even though
prospects for dramatic successes may be much less likely under real
life conditions.

Linking agricultural research to the needs and circumstances
of the rural majority is a high priority for the future and will
probably require important institutional reforms.[3] An important
objective is to identify the priority needs of groups of small
farmers; the most effective means is to enable small farmers to
participate in the research process by expressing their needs,
defining the problems on which researchers should work, and
assisting in testing and evaluating hypotheses and results. How
best to involve small farmers in the research process is not clear
and will no doubt vary from place to place. Among the experiments
now underway is one at ICTA (Institute of Agricultural Science and

---

[3]Professors William F. Whyte and Damon Boynton have organized an
interdisciplinary group of plant, animal and social scientists
under the auspices of the Rural Development Committee of the Cornell Center for International Studies. They are attempting to
evolve organizational patterns that can facilitate the involvement
of small farmers in agricultural research. Publication of their
findings is expected in 1980.

Technology) in Guatemala where small farmers participate actively
in all phases of the research enterprise. The involvement of small
farmers in research is facilitated, as in Taiwan, by farmers'
organizations. The individual small and marginal farmer is poorly
equipped to interact with the agencies of the state, including
research agencies, or to make effective claims on governments.
Organized farmers, however, can express their collective interests
and needs and their organizations can provide the channel through
which small farmers as a group can contribute and exert influence
over research activities, if research organizations and the governments that finance them are inclined to work in that mode. Successful agriculture usually involves farmer inputs into the research
process and this is no less likely to be true of rural development
strategies oriented to the productivity and employment of small
farmers and landless workers. As a practical matter, the latter
can participate only through their organizations.

One way of indirectly involving small farmers, even in the
absence of effective organizations, is through the participation
of social and behavorial scientists in the research enterprise.
The special skills of the social scientists are in the examination
and understanding of social structures and institutional behavior.
Every technological innovation must be introduced into an ongoing
system of production, communication, exchange and power. The more
that is known about these social systems, (e.g. sources and terms
of credit, patterns of labor allocation, sequences of cropping,
channels of marketing, local organizations and structures of
authority, relations with agencies of government, land tenure
arrangements, etc.), the less the risk that research efforts will
be wasted in innovations that the society will not tolerate or which
would cause severe conflict, disruption or even suffering. The
more that is known about real life conditions, the more likely
that institutional and behavioral changes required by more productive
technologies can be guided with minimal disruption, that the
responses of the local community can be monitored, and that timely
adjustments can be made.

Social research can considerably enhance the effectiveness
and the relevance of biological research. This belated recognition
is reflected in the activities of a few of the leading agricultural
research institutions, including the International Rice Research
Institute.[4] The participation of social scientists cannot, however,
be merely symbolic. It will not be sufficient to have a single
house economist in a research group of 100 biologists, or a part-
time consulting anthropologist who is rushed in at the last minute

---

[4]The commitment of IRRI to social research and its increasing contribution to IRRI's decision making are reflected in Research
Highlights for 1977, its most recent annual report, pp.84-87.

to rescue a program from disaster when farmers unexpectedly reject
an innovation in which the agency has invested years of effort.
There must be a critical mass of social scientists building their
knowledge base cumulatively, over time, in exactly the same way that
plant and animal scientists do, and the social scientists must be
integrated into the research enterprise from the very beginning,
including the initial definition of research problems. The participation of social scientists in agricultural research enterprises
can contribute to the relevance of scientific and technological
work and reduce the risk of misdirected effort. It can complement,
but not substitute for greater involvement by organizations representing small farmers.

In our analyses of the relationship of agricultural research
to rural development, especially to the needs of the majority of
poorly endowed farmers, tenants and laborers, we cannot afford to
be naive. The implementation of small farmer strategies, the
achievement of higher productivity, increased employment and dis-
tributional equity involve not only knowledge but also power.
Many of the problems of rural development are structural and these
inefficient and inequitable structures are deeply institutionalized.
Necessary reforms require political action and cannot be accomplished
by knowledge alone. Agricultural research even with ample participation by social scientists, must in the short run work within
the possibilities and constraints afforded by the political economy
of the country concerned, some of which however, are more flexible
and adaptable than others. Thus there can be no simple or instant
social answers. The problems that social scientists deal with are
as hard and intractable in their way as those confronting biologists.

In this brief review, I have identified four challenges that
deserve a prominent place in the agenda of those concerned with
relating agricultural research to rural development: the first is
the continuing priority that must be accorded to building effective
national research institutions in developing countries, institutions
commited to the needs of their own rural publics. They are indispensable to any sustained progress in rural development, and it is
only through them that western scientific institutions can contribute to improving agricultural production and to broadly based
rural development in the many distinctive micro-environments in
which small farmers earn their livelihoods.

The second is the need to orient agricultural research much
more directly than in the past to the farming systems of the majority
of small and marginal cultivators, and to the urgent needs simul-
taneously to increase output per unit of land and to create
additional employment. Rural development will require much greater
research support than has been available in the past, but the
research must be oriented to new sets of problems and relate directly
to new groups of users.

The third is to involve small and marginal farmers in the phases of research, principally their own organizations. This should include defining research problems, contributing information, and testing improved practices.

The fourth is to incorporate social and behavioral scientists into more effective partnership with biological scientists in the joint quest for knowledge and practices that can contribute to broadly based rural development. The intellectual support for rural development must be a broad interdisciplinary effort; it cannot and should not be the burden of the biological scientists alone.

CONTRIBUTORS

Nyle C. Brady
Director General
International Rice Research Institute
Los Banos, Laguna
Philippines

Peter S. Carlson
Department of Crop and Soil Sciences
Michigan State University
East Lansing, MI   48824

John K. Coulter
Scientific Adviser
Consultative Group on International Agricultural
  Research
Secretariat
World Bank
Washington, DC   20433

Milton J. Esman
John S. Knight Professor of International Studies
Cornell University
Ithaca, NY   14853

Ralph W. Hardy
Central Research and Development Department
E.I. DuPont de Nemours & Company
Wilmington, DE   19898

H.T. Huang
National Science Foundation
Washington, DC   20550

C.B. Huffaker
Division of Biological Control
Department of Entomological Sciences
University of California
Berkeley, CA 94720

Paul J. Kramer
Department of Botany
Duke University
Durham, NC 27706

Marvin R. Lamborg
Charles R. Kettering Research Laboratory
150 East South College Street
Yellow Springs, OH 45378

L.G. Mayfield
National Science Foundation
Washington, DC 20550

Thomas R. Odhiambo
Director
The International Centre of Insect Physiology
  and Ecology
Nairobi, Kenya

D. Woods Thomas
Executive Director
Board of International Food and Agricultural
  Development
Washington, DC 20523

Ruben L. Villareal
Asian Vegetable Research and Development Center
Tainan, Taiwan
Republic of China

Israel Zelitch
Department of Biochemistry
Connecticut Agricultural Experiment Station
New Haven, CT 06504

Author Index

Ableson, P.H., 34
Abu-Shakra, S.S., 124, 125, 132, 133
Adkisson, P.L., 174, 175, 194
Agbigay, F.P., 9
Akpan, E.E.J., 56, 61
Alam, M.M., 186, 194
Albrecht, S.I., 148, 150
Alimagno, B.V., 136
Allen, B.R., 88, 99
Andersen, K., 123, 132, 136
Anderson, E.V., 89, 99
Anderson, N., 100
Andres, L.A., 182, 184, 194
Apple, J.L., 172
Arndt, C.H., 56, 61
Atkins, C.A., 134, 135
Aziz, S., 12, 32

Bajaj, Y.P.S., 76
Baker, L., 71, 76
Balaoing, V.G., 9
Balch, R.E., 183, 195
Banfalvi, Z., 134
Barber, L.E., 139, 150
Baskin, C.C., 55, 61
Baskin, J.M., 55, 61
Bassham, J.A., 55, 61
Bauer, W.D., 122, 132, 133
Beaudette, P.D., 113
Bean, E.W., 56, 61
Bednarski, M.A., 135
Begg, J.E., 57, 61
Beglyarov, G.A., 188, 195
Behar, M., 161, 171
Behki, R., 65, 76

Behreus, W.W., III, 135
Bemis, W.P., 97, 100
Bender, R.A., 134
Benjamin, M.R., 81, 98
Bennett, F.D., 186, 192, 194, 195
Benyon, J.L., 133, 134
Bergersen, F.J., 121, 132
Beringer, J.E., 122, 133, 134
Berja, N.S., 136
Berlyn, M.B., 105, 106, 107, 108, 113
Bess, H.A., 179, 195
Bethlenfalvay, G.J., 124, 133
Bhojwani, S.S., 65, 76
Bhuvaneswari, T.V., 122, 132, 133
Bieleski, R.L., 62
Bird, F.T., 183, 195
Biswas, M.R., 156, 171
Black, C.C., 54, 61
Blair, G.J., 21, 33
Boerma, A.H., 141, 142, 150
Bohlool, B.B., 121, 133
Bornemisza, E., 34
Bottino, P.J., 65, 76
Bowers, W.S., 169, 171
Boyer, J.S., 57, 61
Boynton, D., 213
Bradfield, R., 14, 33
Brady, N.C., 11
Bragg, D., 91, 100
Brenchley, J.E., 134
Brill, W.J., 122, 126, 133, 135
Brown, A.W.A., 63, 76
Brown, L.R., 11, 12, 33
Buchanan-Wollaston, A.V., 133, 134

Bulen, W.A., 136
Buol, S.W., 21, 34
Burges, H.D., 195
Burgess, B.K., 135
Burns, A., 136
Burris, R.H., 6, 54, 61, 121, 134
Byerly, T.C., 76

Callaham, D., 126, 133
Calvert, H.E., 126, 134
Calvo, F., 34
Campbell, N.E.R., 134
Cannell, M.G.R., 61
Carl, K.P., 186, 194
Carlson, P.S., 63, 65, 76
Carrillo S., J.L., 197
Carter, M.C., 54, 61
Catedral, I.C., 9
Caughley, G., 182, 195
Chandler, R.F., 8
Chang, T.T., 22, 33
Chelliah, S., 171
Chen, N.C., 76
Chien-kang, C., 76
Child, J.J., 135
Ching, T.M., 118, 120, 133
Christiansen, M.N., 56, 62
Chun-Chow, T., 76
Clausen, C.P., 23, 185, 195
Cleere, J., 171
Cochereau, P., 195
Cocking, E.C., 28, 33, 76
Comroe, J.H., Jr., 81, 98
Coppedge, J.R., 197
Coulter, J.K., 35
Covarrubias G.,R., 197
Crawford, D.L., 88, 99
Cresswell, M.M., 62
Crist, D., 118
Crookston, R.K., 56, 61
Cummings, R.W., Jr., 4, 8, 34,
    109, 114

Dalrymple, D.G., 15, 33
Dancel, W.A., 9
Darrow, R.A., 118, 133, 135
Dazzo, F.B., 122, 133
DeBach, P., 174, 178, 187, 195,
    196

Del Tredici, P., 133
DeLeo, A.B., 134
Dethier, V.G., 163, 166, 171
Dickerson, W.A., Jr., 197
Dilworth, M.J., 135
Dodd, A.P., 182, 195
Doutt, R.L., 162, 171, 178, 195
Dripps, R.D., 81, 98
Duncan, W.G., 54, 55, 61

Earle, E.D., 8
Earp, D.A., 198
Eckholm, E.P., 11, 33, 89, 99
Eligio, D.T., 9
Elkan, G.H., 121, 136
Elliott, J., 135
Emerich, D.W., 150
Esman, M.J., 209, 211
Espinas, C.R., 136
Evans, H.J., 118, 124, 133, 134,
    139, 150
Evans, J.J., 116, 136
Evans, L.T., 53, 54, 55, 57, 61
Evans, P.K., 76
Evenson, R.E., 109, 110, 111, 113

Ferguson, A.R., 62
Ferrar, P.J., 56
Ferrari, T.E., 6, 8
Fishback, J., 83, 98
Flint, E.P., 55, 56, 62
Ford, G.R., 115
Forrai, T., 134
Foster, K., 91, 100
Francis, C.A., 34
Franz, J.M., 184, 195
Frazer, B.D., 198
Fujihara, S., 121, 133
Fulkerson, J., 83, 98

Garcia, R.L., 142, 150
Gardner, I.C., 121, 133
Giaquinta, R.T., 144, 150
Gibbs, M., 76, 114
Gibson, A.H., 135
Gifford, R.M., 55, 61
Ginsburg, A., 117, 133
Gonzalez, D., 186
Gracen, V.E., 8

Gregory, P., 5, 8
Guyford Stever, H., 80, 98

Hadfield, K.L., 136
Hagen, K.S., 186, 193, 194, 195
Hall, P.L., 88, 99
Hamai, J., 175, 176, 196
Hammond, A.L., 89, 99
Han, H., 69, 76
Hanson, G., 91, 93, 99, 100
Hanus, F.J., 134, 150
Hanus, J., 133
Hanway, J.J., 142, 150
Haramoto, F.H., 195
Hardy, R.W.F., 103, 114, 123,
    124, 133, 134, 137, 138, 140,
    141, 142, 144, 147, 150, 151
Hargrove, T.R., 28, 33
Harper, J.L., 182, 195
Harwood, R.R., 159, 160, 161
Hassell, M.P., 176, 179, 180,
    181, 195
Havelka, U.D., 103, 114, 123,
    124, 133, 134, 137, 141, 142,
    144, 147, 151
Hedtke, S., 133
Heinrichs, E.A., 166, 167, 171
Heinz, D.J., 69, 76
Helms, J.A., 54, 61
Hendriksen, A.J.T., 33
Herridge, D.F., 121, 134
Hesketh, J.D., 54, 55, 61
Heytler, P.G., 147, 151
Hirsch, P.R., 134
Hogan, L., 96, 100
Hoggan, S.A., 133
Hollaender, A., 132, 133, 135,
    136, 150
Holzer, H., 136
Hom, R., 198
Hooykaas, P.J.J., 122, 134
Horecker, B.L., 134
Howeler, R.H., 34
Hsiao, T.C., 57, 61
Huang, H.T., 79
Hubbell, D.H., 122, 133
Huber, D.M., 139, 151
Huffaker, C.B., 162, 171, 173,
    174, 176, 177, 178, 183, 184,
    185, 188, 190, 195, 197, 198

Huffaker, R.C., 132
Huffman, K.W., 62
Hussey, N.W., 191, 192, 195, 196
Hutauruk, C., 192, 196

Isenson, R.S., 98
Isrankura, V., 20, 33

Jacobs, T., 76
Jennings, N., 133
Jennings, P.R., 15, 33
Jensen, L., 83, 98
Jiminez J., E., 188, 196
Johnson, S.W., 110
Johnston, A.W.B., 122, 123, 133,
    134

Keister, D.L., 116, 134, 135
Kennedy, G.G., 163, 164, 167, 171
Kennett, C.E., 174, 184, 190, 196
King, P.J., 77
Kiss, G.B., 134
Klapwijk, P.M., 134
Kleinhofs, A., 65, 76
Klucas, R.V., 121, 134
Knight, C.G., 13, 33
Knipling, E.F., 162
Knotts, R.R., 118, 133
Kogan, M., 168, 171
Kok, B., 55
Kondorosi, A., 122, 134
Kozlowski, T.T., 61
Kramer, P.J., 51
Krishnawurth, M., 76
Kurz, W.G.W., 116, 134

Lacewell, R.D., 192
Ladisch, C.M., 88, 99
Ladisch, M.R., 88, 99
Laing, J.E., 175, 176, 196
Lalonde, M., 126, 134
Lamborg, M.R., 115, 139, 144
Lantican, R.M., 5, 9
LaRue, T.A., 116, 134
Last, F.T., 61
Latzko, E., 114
Lawyer, A.L., 105, 114
Leaf, G., 121, 133
Ledeboer, A.M., 135
Lee, R., 61

Lenz, F., 54, 61
Lepidi, A.A., 135
Lepkowski, W., 83, 98
Lewis, O.A.M., 121, 134
Luck, R.F., 196
Luckmann, W.H., 171

Magasanik, B., 117, 134
Magate, L.Z., 9
Maggs, D.H., 54, 62
Maier, R.J., 134, 150
Mamaril, C.P., 33
Mansfield, T.A., 169, 171
Maretzki, A., 76
Marini-Bettolo, G.B., 172
Markkula, M., 192, 196
Marsella, P., 171
Martineau, R., 197
Matsumoto, T., 121, 134, 135
Matteson, P., 198
Matthews, D.E., 8
May, R.M., 195
Mayfield, L.G., 79
Mayne, B.C., 135
McCombs, J.A., 116, 135
McIntyre, D., 91, 99
McMurtry, J.A., 174, 196
McWilliam, J.P., 56, 62
Meade, H.M., 122, 135
Meadows, D.H., 115, 135
Meadows, D.L., 135
Messenger, P.S., 185, 195, 196, 197
Metcalf, R.L., 171
Miller, W.P., 98, 100
Milner, M., 151
Mitsui, A., 114, 151
Miyashi, S., 151
Momat, E., 33
Monge C., J., 197
Moore, A.W., 129, 135
Moreira, T.J.S., 171
Morris, R.F., 183, 197
Morrison, R.K., 197
Mort, A.J., 132
Morton, J.A., 80, 98
Mothes, K., 121, 135
Munger, H.M., 5, 7
Murashige, T., 91, 99

Nagatani, H.H., 126, 135
Neilson, M.M., 183, 184, 197
Nelson, D.W., 151
Nesbit, D.P., 192, 198
Newcomb, W., 133
Newton, W.E., 126, 135
Nickell, L.G., 76
Nieminen, M., 196
Norris, D.M., 88, 99
Nowierski, R.M., 186
Nuti, M.P., 122, 134, 135
Nutman, P., 134

Odhiambo, T.R., 153, 157, 162, 171
O'Gara, F., 121, 135, 136
Ogren, W.L., 55
Ohta, T., 171
Oka, H.I., 22, 33
Okigbo, B.N., 159, 160, 171
Oliver, D.J., 55, 62, 105, 106, 107, 114
O'Neill, D.J., 89, 99
Orton, T., 71, 76
Osmond, C.B., 55
O'Toole, J., 61
Ozbun, J.L., 61

Pacheco M., F., 188, 197
Pagen, J.D., 116, 135
Painter, R.H., 164, 171
Pang, T.C., 192, 197
Paris, C.G., 134
Parker, F.D., 190, 197
Parrot, J., 76
Pate, J.S., 121, 134, 135
Pathak, M.D., 165, 167, 171
Patterson, D.T., 55, 56, 62
Peters, G.A., 127, 128, 135
Phillips, D.A., 132, 133
Phillips, J.R., 193
Pimentel, D., 186, 197
Polacco, J.C., 65, 76
Ponnamperuma, F.N., 21, 34
Porter, B.A., 174, 197
Potrykus, I., 77
Powell, R.D., 62
Power, J.B., 28, 33
Powles, S.B., 55, 62
Prival, M.J., 134

Prusiner, S.M., 133, 136
Pueppke, S.G., 133

Rabb, R.L., 187, 197
Radmer, R., 55, 62
Rainbird, R.M., 134
Rains, D.W., 129, 131, 136
Randers, J., 135
Rao, V.R., 118, 135
Raper, C.D., Jr., 56, 62
Ray, T.B., 135
Reilly, P.J., 88, 98
Reinert, J., 76
Reporter, M., 116, 135, 136
Reynolds, H.T., 174, 175, 197
Rick, C.M., 4, 8
Ridgway, R.L., 187, 195, 196, 197, 198
Ripper, W.E., 174, 197
Rodriguez, E., 91, 99
Rolfe, B., 136
Rörsch, A., 134
Rose, A.H., 197
Rosen, D., 195
Rowley, J.A., 56
Rubis, D., 91, 99
Ruiz-Arqüeso, T., 133
Russell, R.S., 58, 62
Russell, S.A., 134, 150
Ruttan, V.W., 111, 113

Salt, G., 186, 197
Sanchez, P.A., 21, 34
Sankaran, T., 192, 197
San Pietro, A., 76, 151
Sarkanen, K.V., 88, 99
Saxena, R.C., 165, 166, 167, 171
Schilperoort, R.A., 134, 135
Schmidt, E.L., 121, 133
Schultz, T.W., 157, 172
Schuster, M.F., 195
Schutt, H., 136
Scopes, N.E.A., 191, 192, 196
Scowcroft, W.R., 28, 34, 135
Scrimshaw, N.S., 151, 161, 171
Servaites, J.C., 55, 62
Setchell, A.M., 134
Shafizadeh, F., 88, 99
Shah, V.K., 126, 135

Shanmugan, K.T., 121, 132, 135, 136, 139, 151
Sharkey, P.J., 135
Sherwin, C.W., 81, 98
Siddall, J.D., 169, 172
Signer, E.R., 122, 134
Simpson, F.B., 134
Situmorang, H.S., 192, 196
Skotnicki, M., 116, 136
Smetnik, A.I., 188
Smith, R.F., 162, 171, 172, 174, 175, 197
Sneep, J., 33
Sogawa, K., 29
Spain, J.M., 21, 34
St. John, J.B., 56, 62
Stadtman, E.R., 117, 133, 134, 136
Stevens, M.A., 8
Stiefel, E.I., 135
Stinner, R.E., 188, 197
Storey, R., 116, 136
Streeter, J.G., 121, 136
Streicher, S.L., 134
Swidt, J.E., 194
Syed, R.A., 192, 197

Tai, W., 76
Takase, K., 34
Talley, B.S., 129, 131, 136
Talley, S.N., 129, 131, 136
Tamura, S., 151
Taylor, A.O., 56
Thomas, D.W., 199
Thomas, E., 65, 77
Thomas, J.F., 56, 62
Thorpe, T.A., 76
Tiittanen, K., 196
Tjepkema, J.D., 116, 136
Toia, R.E., Jr., 135
Torrey, J.G., 133
Tsai, C.Y., 151
Tsao, G.T., 88, 99
Tsun-wen, O., 76
Tubbs, R.S., 121, 136
Turgeon, G., 132
Turner, G.L., 121, 132
Turner, N.C., 57, 61
Tyler, B.M., 134
Tze-ying, H., 76

Upchurch, R.G., 121, 136

Valentine, R.C., 123, 132, 136,
  139, 151
van den Bosch, R., 186, 195,
  197, 198
van den Vrie, M., 196
Verbiscar, A.J., 96, 100
Viktorov, G.A., 195
Villareal, R.L., 1, 3, 5, 7, 9
Vincze, E., 134
Vinson, S.B., 187, 189, 190, 195,
  196, 197, 198

Wade, N., 59, 62
Waggoner, P.E., 111, 113
Wallace, D.H., 6, 8, 61
Walter, J.M., 9
Wang, D.I.C., 151
Warren, H.L., 151
Wastie, R.L., 196, 198
Watanabe, I., 129, 136
Watson, D.J., 53, 54, 55, 62
Watt, G.D., 123, 136
Weiss, C., Jr., 109, 114

Wellburn, A.R., 171
Whyte, W.F., 213
Wickham, T., 34
Wilcox, R.P., 13, 33
Wilhelm, S., 158, 159, 172
Williams, C.M., 153, 172
Williams, C.N., 54, 61
Wohlhueter, R.M., 117, 136
Wood, B.J., 174, 192, 195, 196,
  198
Wortman, S., 34
Wortman, W., 109, 114

Yamaguchi, M., 133
Yamamoto, Y., 121, 133, 134, 135
Yasumatsu, K., 195
Yatazawa, M., 134, 135
Yermanos, D.M., 96, 100
Yokoyama, H., 92, 93, 100
York, D.W., 8
Yoshida, S., 22, 34

Zelitch, I., 55, 62, 85, 101, 102,
  103, 104, 105, 106, 108, 113, 114

Subject Index

Abscisic acid, insect resistance
    and, 169
Acetylene, reduction of, by
    Rhizobium japonicum, 117
Actinomycetes, lignin degradation
    by, 88
Africa
  agricultural production increases,
    in, 156
  malnutrition in, 13
Agasicles hygrophila Selman and
    Vogt, as plant pre-
    dator, 182
Agricultural Experiment Stations,
    advantages of, 113
Agricultural research
  in developing countries, 1-4
    rate of return from, 108-109
  funding for, 81-84
  organization for effectiveness
    of, 110-113
  rural development and, 209-216
Agricultural traits, genetics
    of, 67
Agrobacterium sp., nitrogen
    fixation by, 149
Agrobacterium tumefaciens, plasmid
    of, 122
Alcohol, from sugar cane and
    cassava, 89
Alfalfa
  insect pest resistance of, 165
  photorespiration in, 104
Alfalfa weevils, biological
    control of, 185,186

Algae, as rice nitrogen
    source, 23, 31
Allantoic acid and allantoin,
    from soybean nitrogen
    fixation, 121
Allelomic factors, role in
    insect resistance, 168
Alligator weed
  insect predators of, 182
    economic benefits from, 184
Allomones, role in insect
    resistance, 168
Alnus, nitrogen fixation in, 126
Alternanthera philoxeroides
    (Mart.) Griseb.
    see Alligator weed
Aluminum
  rice sensitivity to, 21
  wheat tolerance to, 46
Ammonia, world production
    of, 138
Amphorophora agathonica,
    resistance to, 163, 166
Anabaena-azolla, as part of
    nitrogen-fixation
    system, 126-129, 139
Aonidiella auranti, resistance
    to, 163
Apanteles sp., biological
    control by, 186, 194
Aphytis maculicornis (Masi),
    in olive scale con-
    trol, 184-185
Apples
  insect pest resistance of,
    165, 166

Apples (cont'd)
  yield of, photosynthesis and, 54
Arid land, utilization of plants
          from, 89–98
Asia
  fertilizer consumption in, 21
  food problems of, 12–13
Asian Vegetable Research and
          Development Center, 3, 6
Asparagine, transport function,
          in lupines, 121
ATP, requirement of, in nitrogen
          fixation, 123, 148
Azolla
  as green manure, 129
  as part of nitrogen-fixation
          system, 126–129, 139, 149
      for rice, 23, 31, 129–131, 139,
          149
Azospirillum-cereal grain,
          nitrogen fixation in, 139
Azotobacter vinelandii
  glutamine synthetase in, 118
  nitrogenase in, 126

Bacterial blight, rice screening
          for resistance to, 25
Bacteroids, experimentally
          induced, 118
Bagworms
  biological control of, 192
  pesticide-induced, 174
Banana Research Institute, 4
Barley
  crop improvement of, 48
    genetics in, 70
  insect pest resistance of, 165
  international tests on, 40
  major diseases and pests of, 45
Basic research
  on crop improvements, 1–9
  time lag to benefits from, 80
Bathyplectes sp., use in weevil
          control, 186
Beans   (See also Legumes)
  international tests on, 40
  weather effects on, 56
Behavioral chemicals, use in
          biological control, 189–
          190

Biological control
  of insect pests, 173–198
    chemical control combined
          with, 190–194
    empirical basis, 175–177
    theoretical basis, 177–181
Biomass
  basic research on, 101–114
  fuels from, 88–89
  production and utilization
          of, 79–100
Black gram, see Vigna mungo
Blast fungus
  rice resistance to, 44
    screening, 25, 27
  strains of, 30
Bollworms
  biological control of, 189,
          192–193
  pesticide-induced, 174
Brassicas, self compatibility
          in, 6
Brazil, biomass conversion in,
          89
Broccoli, self-compatibility
          studies on, 6
Brown planthopper
  importance as pest, 44
  pathogens affecting, 30
  resistance to, 29, 166–168
Budworms
  biological control of, 192
  pesticide-induced, 174
Buffalo gourd, research on,
          97–98

C_3 carbon pathway, efficiency
          of, 55
C_3 plants, photorespiration
          in, 103
  rates, 104
C_4 carbon pathway, efficiency
          of, 55
C_4 plants, photorespiration
          in, 103
  rates, 104
Cabbage
  KK hybrid of, 7
  self-compatibility studies
          on, 6

Cactoblastis cactorum Berg,
      as Opuntia predator, 182
      187
Carbon dioxide, biomass pro-
      duction from, 101
Carotenoids, biosynthesis of, 92
Cassava
  alcohol from, 89
  genetic resource conservation,
     of, 41
  international tests on, 40
  land use for, 39
  major diseases and pests of, 45
  wheat replacement of, 48
  world yield of, 115
Cassava mealy bug, spread of, 39
Caterpillars, biological control
    of, 192
Cell, manipulations of, in crop
    improvement, 66
Cell systems, use in plant
    improvement studies, 68-
    71
Cellulose
  hydrolysis of, 88
  in wood, 86
Cercopid pests, biological con-
    trol of, 193
Cereals, production in developing
    countries, 3, 154
Chaetosiphon fragaefolii, resis-
    tance to, 163
Chestnut, insect pest resistance
    of, 165
Chestnut blight organism,
    accidental introduction
    of, 187
Chick pea
  genetic resource conservation
    of, 40
  international tests on, 40
  major diseases and pests of, 45
Chilo suppresalis, see Stem
    borer
China, agricultural research
    in, 1-2
Chinese cabbage, heat tolerance
    screening of, 6

Chlorinated phenoxy
    triethylamines, bio-
    regulatory activity
    of, 93
Chromaphis juglandicola (Kalt.),
    see Walnut aphid
Chrysolina quadrigemina Suffr.,
    see Leaf-feeding beetle
Citrulline, transport function,
    in alders, 121
Citrus fruits, insect pest
    resistance of, 163
Climate, adverse, plant breeding
    for, 46-47
Clover, insect pest resistance
    in, 165
Coccophagoides utilis Doutt,
    in olive scale control,
    184-185
Comptonia peregrina, 126
Consultative Group on Inter-
    national Agricultural
    Research, 37
  work of, 38
Corn. (See also Maize)
  opaque-2 mutant gene of, 6
  susceptibility to H. maydis,
    5
Cotton
  insect pests of
    management, 192-193
    resistance, 165
  land use for, 39
  weather effects on, 56
Cottony cushion scale, insect
    control of, 174
Cowpea
  genetic resource conservation
    of, 40
  international tests on, 40
  major diseases and pests
    of, 45
Crop production
  constraints to, removal, 11-
    34
  genetics use in, 39-41, 63-77
  improvement of, 35-49
    at international centers,
    41-44

Crop production (cont'd)
    physiology role, 51-62
    for rice, 14-17
  insect control and, 153-172
  mixed cropping in, 158-162
  photosynthesis and, 54-55
    factors affecting, 103
  technology development for, 38-
      39
  weather importance in, 55-56
Crops
  basic research on, 4-6
  improvement programs for, in
      developing countries, 1-9
Cucumber beetle, resistance to,
    164-165
Cucurbita foetidissima see
      Buffalo gourd
Cucurbitaceae, insect pest resis-
      tance in, 164, 165
Cucurbitcacins, role in insect
      resistance, 164, 165
Cyclamen mites
  biological control of, 190
  pesticide-induced, 174

Dacus dorsalis Hendel, see
      Oriental fruitfly
DDT, contraindication for insect
      control, 174
Defoliation, effect on plant
      physiology, 51
Developing countries
  agricultural research in, 1-4,
      80
    rate of return from, 108-109
    rural development and, 209-216
  biological nitrogen fixation
      in, 115-136
  crop productivity in, 3
    biological nitrogen fixation
        and, 137-151
  fertilizer consumption in, 21
  meat production in, 142
  plant genetic role in, 64
  population growth in, 155
  staple crops of, 154
Diatraea saccharalis (F.) see
      Sugarcane moth borer

Dieldrin, selective use of, 192
Diprion hercyniae, see
      European spruce sawfly
DNA, manipulation of, in crop
      improvement, 66, 68
Downy mildew
  in mixed crops, 160
  muskmelon resistant to, 7
Drino bohemica (Mesn.), as
      parasitoid, 184
Drought
  effect on
    plant physiology, 51
    rice production, 19, 30-31
  tolerance to, 58

Egypt, agricultural research
      in, 2
Egyptian alfalfa weevil, bio-
      logical control of, 186
Emblema amabilis, biological
      control of, 190
Encarsia formosa Gahan, bio-
      logical control of, 191
Endothia parasitica see
      Chestnut blight organism
Environment, effect on plant
      physiology, 52
Enzymes, in hybrids, 73
2,3-Epoxypropionate, see
      Glycidate
Erwinia sp., nitrogen fixation
      by, 149
European corn-borer see
      Ostrinia nubilalis
European spruce sawfly, bio-
      logical control of, 183
Exenteron vellicatus Cush., as
      parasitoid, 184

Far East, malnutrition in, 13
Farming systems, need to
      improve, 212-213
Farnesol, insect resistance and,
      169
Fertilizer(s)
  biological nitrogen fixation
      as, 115-136
  effect on rice production, 23

Fertilizers (cont'd)
  irrigation and, 48
Fescue, photorespiration in, 104
Field beans, major diseases
        and pests of, 45
Filberts, insect pest resistance
        of, 163
Floods
  effects on rice production,
        18-19
    resistance, 32
Floriculture, meristem and shoot
        tip culture in, 69
Food
  future deficits of, 11
  improved production of, 36
  sources, 13-14
Fuels, from biomass, 79-100
Funding, for agricultural R+D,
        81-84

Gasohol, 89
Genetics
  of agricultural traits, 67
    desirability, 72-75
  in crop improvement, 4-5,
        39-41, 63-77
    rice, 28-29
  effects on plant physiology,
        52
  manipulation of, 65
  use in plant breeding, 39-43
Germ plasm
  basic forms of, 39-40
  collection for major crops, 41
    rice, 24
Glasshouse pests, biological
        control of, 191-192
Glutamate:glyoxylate aminotrans-
        ferase pathway, glycidate
        inhibition of, 105
Glutamine synthetase
  in rhizobia, 117-118
  properties, 118
Glycidate, as photorespiration
        inhibitor, 104-105
Glycolate pathway, of photo-
        respiration, 103-106
Graduate students, in plant
        physiology, research
        philosophy of, 59

Grains
  biological nitrogen fixation
        for, 139
  hybrids of, 43
  Mexican trade changes in, 11-12
  wheat replacement of, 48
  world production of, 138
Grants, competitive, disadvantages
        of, 111-112
Grasses, weather effects on, 56
Grassy stunt disease
  rice resistance to, 44
    screening for, 25
Green leafhopper, see
        Nephotettix virescens
        (Distant)
Groundnut, see Peanut
Guayule
  commercialization of, 48
    research on, 90-95
Gulfstream muskmelon, 7

Haploid selection method, in crop
        improvement, 69
Heliothis spp., biological control
        of, 188
Helminthosporium maydis, sus-
        ceptibility of, genetics,
        5
Hemicellulose, hydrolysis of,
        87-88
Heterosis, characterization of,
        72-75
Hevea rubber, guayule rubber as
        substitute for, 90
Hippodamia convergens Guer.-Men.,
        use in biological control,
        187
Hordeum jubatum, interbreeding
        with H. vulgare, 70-71
Hordeum vulgare   (See also
        Barley)
  genetics in improvement of, 70
Hybridization, of rice, 24, 25
Hypera brunneipennis (Boh.), see
        Egyptian alfalfa weevil

Indian lands (U.S.), jojoba
        development on, 97-98
Insect pests
  control of, 153-172

Insect pests (cont'd)
    approaches to, 162
  crop resistance to, 42, 44-46
  effect on rice production,
        9-10
    resistance factors, 29-30
  pesticide induced, 174
International Agricultural
        Research Centers, 209-
        210
International Board for Plant
        Genetic Resources, 40
International Maize and Wheat
        Improvement Center, 3,
        37
International Rice Research
        Institute, 3, 37
Iron
  in soil, effect on rice, 21,31
  wheat tolerance to, 46
Iron-molybdenum cofactor, of
        nitrogenase, 116, 124-
        126
Isonicotinic acid hydrazide
        (INH), effects on glyco-
        late metabolism, 107-108

Japan, agricultural research
        in, 2
Jojoba
  commercialization of, 48
    research on, 95-96
  oil from, 95
  seed meal, amino acid compo-
        sition of, 96
  toxic compounds in, 96
Juvenile hormones, insect resis-
        tance and, 169

Kairomones
  biological control by, 189-190
  role in insect resistance, 168
KK cabbage hybrid, 7
Klamath weed, see St. Johnswort
Klebsiella pneumoniae
  glutamine synthetase in, 118
  nitrogen fixation in, 123

Ladybird beetle, see Hippodamia

Latin America, malnutrition in,
        13
Leaf
  expansion of, importance in
        photosynthesis, 55
    water effects, 57
Leaf-feeding beetle, as St.
        Johnswort predator, 182
Leafhoppers, rice screening for
        resistance to, 25
Lectins, of legumes, rhizobia
        binding by, 121
Legumes
  future needs for, 140-142
  genetic resource conservation
        of, 41
  genetics in improvement of, 71
  hybrids of, 43
  nitrogen fixation in, 124, 144,
        147-148
  production in developing
        countries, 3
  recognition/infection process
        in, 121-122
Less developed countries, see
        Developing countries
Levoglucosan, from lignocellulose,
        88
Lignin, degradation by actino-
        mycetes, 88
Lignocellulose, conversion to
        feedstocks, 87
Lignocellulosic residues,
        generation of, in U.S.,
        85
Lycopenes, rubber biosynthesis
        and, 92-93
Lycopersicon, disease resistance
        of, 5

Maize   (See also Corn)
  genetic resource conservation
        on, 40-41
  hybrids of, 43
  insect pest resistance of, 165
  international tests on, 40
  major diseases and pests of, 45
  photorespiration in, 104, 105
  yield of photosynthesis and, 54

Maize (cont'd)
  world yield of, 115
Maize stem-borers, in mixed
        crops, 160-161
Malaysian Rubber Institute, 3
Malnutrition, estimated areas
        of, 13
Meat, world production of, 142
Meristem culture, value in
        floriculture, 69
Metarrhizium anisopliae, use in
        borer control, 193-194
Mexico
  agricultural research in, 2
  grain trade changes in, 11-12
Microbracon greeni Ash., bio-
        logical control by, 190
Millet
  major diseases and pests of,
        45
  pearl variety, see Pearl
        millet
Mites
  biological control of, 189
  predatory, use in mite con-
        trol, 189, 191
Mixed cropping
  entomology of, 158-162
    importance, 170
  pest management in, 163
Mudgo rice variety, planthopper
        resistance of, 29
Mung bean, hybrid studies on, 71
Muskmelon, downy mildew-
        resistant, 7
Mutants, photorespiration
        alteration in, 106-108
Myzus percicae, biological con-
        trol of, 191

Near East, malnutrition in, 13
Nephotettix virescens, rice
        resistance to, 167
Nilaparvata lugens, see Brown
        planthopper
Nitrate, heterosis and levels
        of, 74
Nitrate reductase, in hybrids,
        73-75

Nitrite reductase, in hybrids, 74
Nitrogen deficiency, effect on
        plant physiology, 51
Nitrogen fixation
  abiological, 139
  advances in research on, 142
    table, 143
  biological, 115-151
    for cereal grains, 139
    crop improvement from, 137-
        151
    problems in, 142, 144, 145-
        147
  future technologies for, 148-
        149
  in legumes, 124, 144, 147-148
  in non-legume plants, 7
  photosynthesis and, 123-124
  by Rhizobium, 116-117
Nitrogenase
  iron-molybdenum cofactor of, 116
    properties, 124-126
  in nitrogen fixation, 147
N-Serve, 139

Oil palms, control of pests on,
        192
Oleoresin, increase of, in pines,
        89
Olive parlatoria scale, biological
        control of, 184
Opius sp., as fruitfly parasitoids,
        179
Opuntia, insect predators of,
        182, 187
Oriental fruitfly, parasitoids
        of, 179
Oryza glaberrima Steud., insect
        resistance of, 167
Oryza nivara, pest resistance of,
        44
Ostrinia nubilalis, resistance
        to, 165, 166
Oxidative phosphorylation, in
        hybrids, 73

Parasitoids, use in biological
        control, 173-198
Parthenium argentatum Gray see
        Guayule

Peanuts
  international tests on, 40
  major diseases and pests of,
      45
Pear, insect pest resistance of,
    163
Pearl millet
  genetic resource conservation
      of, 41
  hybrids of, 43
Peat soil, low productivity for
    rice in, 21
Philippine Sugar Research
    Institute, 3
Phaseolus sp., genetic resource
    conservation of, 40
Photorespiration
  glycolate pathway of, 103–106
  reduction studies on, 55
  role in crop yield, 103
Photosynthesis
  crop yield and, 54–55, 103
  genetic studies of, 6–7
  increasing efficiency of, 55
  nitrogen fixation and, 123–124
  research effort organization
      on, 112
Phycoptella avellanae, resis-
    tance to, 163
Physiology
  environment effects on, 52
  genetic effects on, 52
  limitations in contributions
      of, 53–54
  role in crop improvement, 51–
      62
  suggestions for increasing
      role of, 57–59
  water stress effects on, 56–
      57
  weather effects on, 56
Phytoseiulus persimilis A.-H.,
      use in biological
      control, 191
Pigeon pea
  genetic resource conservation
      of, 40
  international tests on, 40
  major diseases and pests of, 45

Pines, oleoresin increase in, 89
Plant diseases, resistance to,
      44–46
Planthoppers, rice screening for
      resistance to, 25, 27,
      29
Plants
  improvement of, cellular system
      studies, 68–71
  quarantine for, 47
Plasmids, in Rhizobium species,
      122
Polymers, derivation of, 86
Potatoes
  crop improvement of, 48
    genetic aspects, 69
  genetic resource conservation
      on, 40–41
  increasing demand for in
      tropics, 48
  major diseases and pests of, 45
  photorespiration in, 104
  spindle tuber virus of, 47
  world yield of, 115
Precocenes, insect resistance
      and, 169
Predators
  use in biological control, 173–
      198
    examples, 182–187
Prickly pears, see Opuntia sp.
Psylla pyricola, resistance to,
      163
Purple lac culture, biological
      control methods in, 190
Pyricularia oryzae, see Blast
      fungus

Quarantine, of plants, 47
Quinoa, crop improvement in, 48

Raspberry, insect pest resis-
      tance of, 163, 166
Rhizobia
  biological nitrogen fixation
      by, 141–142, 148, 149
  cultures of, nitrogen fixation
      studies on, 120
  glutamine synthetase in, 117–118

Rhizobium japonicum
  hydrogenase in, 124
  lectin binding of, 121-122
  nitrogen fixation in, 116-117
Rhizobium leguminosarum
  linkage maps of, 121
  plasmid transfer to, 123
Rhizobium meliloti, linkage maps
      of, 122
Rhizobium phaseoli, plasmid
      transfer studies on, 123
Rhizobium trifolii, plasmid
      transfer to, 122, 123
Rice   (See also individual
      species)
  as adaptable crop, 13
  Asian food problems and, 11-12
  Azolla as nitrogen source for,
      23, 31, 129-131, 139
  disease resistance of, 25,
      29-30, 44
  dwarf variety of, 5
  fertilizer nitrogen use for,
      139, 140
  as food staple of poor, 13-15
  germ plasm collection of, 24,
      41
  haploid selection method for,
      69
  hybridization of, 24, 25
  insect pest resistance in,
      165-167
  international tests on, 40
  irrigation for, 42
  major diseases and pests of,
      45
  Pokkali variety, resistance
      to adverse soils, 31
  screening of, for stress
      studies, 24
  tolerance to high iron levels,
      46-47
  Tongil variety, disease resis-
      tance, 44
  wheat replacement of, 48
  world yield of, 115
Rice bean, see Vigna umbellata
Rice production
  adverse soil effects on, 20-22

Rice production (cont'd)
  disease effects on, 19-20
    resistance in, 15, 29-30
  double crops in, 15
  drought effects on, 9
    resistance, 30-31
  fertilizer effects on, 23
  genetics in, 28-29
  insect effects on, 19-20
    resistance in, 29-30
  International Rice Testing
      Program for, 25-27
  in Philippines, 37,38
  soil effects on, 20-22, 23,
      31-32
  stresses and constraints on,
      17-22
    list, 18
    removal, 22
    tolerance studies, 23-27
  temperature and climate effects
      on, 14, 22
  water supply and, 18-19, 23,
      30-31, 32, 48
Rodolia cardinalis, see Vedalia
      beetle
Roots, role in rice drought
      resistance, 30
Rubber, from guayule, 90
Rural development, agricultural
      research role in, 209-
      216

Salinity
  effect on rice production, 21
    international tests, 27
  wheat tolerance to, 47
Salvinia, carbon dioxide fixation
      in, 127
Seedlings, low-temperature
      tolerance in, 58
Sheath blight, rice screening for
      resistance to, 25, 27
Shoot tip culture, value in
      floriculture, 69
"Shuttle" breeding, 42
Silviculture plantation,
      experimental, 88-89

Simmondsia chinensis (Link)
          Schneider, see Jojoba
Simmondsin, as jojoba toxic sub-
          stance, 96
Soil, effect on rice production,
          20–22
   management, 23, 31–32
Soil mulch, in rice production,
          23
Sorghum
   crop improvement of, 48
   genetic resource conservation
          of, 41
   hybrids of, 43
   major diseases and pests of,
          45
   weather effects on, 56
Southern corn leaf blight
          disease, host genetics
          and, 5
Soybean(s)
   genetic resource conservation
          of, 40
   lectin function in, 121–122
   nodules, nitrogen fixation in,
          121
   photorespiration in, 104
   weather effects on, 56
   yield of, photosynthesis and,
          54
SP mutant, of tomato, 4–5
Sperm whale oil, jojoba oil
          as substitute for, 95
Spider mite, see Tetranychus
          urticae
Spindle tuber virus, 47
Spurred anoda, 56
St. Johnswort, biological con-
          trol of, 182, 183
Stem borers, rice screening for
          resistance to, 25, 27,
          29
Strawberry, insect pest resis-
          tance of, 163, 164–165
Stresses, on rice production,
          17–24
   tolerance studies, 23–27
Sucrose, role in nitrogen fixa-
          tion, 144

Sugar beet, photorespiration in,
          104
Sugar cane
   alcohol from, 89
   biological control of pests
          in, 193–194
   genetics in improvement of, 69
Sugarcane moth borer, biological
          control of, 186, 193
Sulfur, soil deficiency in rice
          production, 21
Sunflower, photorespiration in,
          104
Sweet potato, major diseases and
          pests of, 45

Teff, crop improvement in, 48
Temperature
   crop yield and, 56
   effect on rice production, 22
      international tests, 27
Tetranychus urticae
   biological control of, 191
   pesticide-induced, 174
   resistance to, 163, 164–165
Tissue culture, virus elimination
          from, 47
Tobacco
   haploid selection in, 69
   insect pest resistance in, 165
   photorespiration in, 104, 105
      mutation effects, 106–107
   weather effects on, 56
Tobacco budworm
   biological control of, 189
   pesticide-induced, 174
Tomatoe(s)
   disease resistant gene transfer
          to, 5
   insect pest resistance in, 165
   self-pruning (SP) character
          in, 4–5
Toxicity, of soil, to rice pro-
          duction, 21–22
Trees, photosynthesis and yield
          of, 54
Trialeurodes vaporariorum, bio-
          logical control of, 191

Trichlorphon, selective use
     of, 192
Trichoderma reesei, cellulases
     from, 88
Trichogramma, use in biological
     control, 187–189, 190
Trioxys pallidus, walnut aphid
     control by, 185–186
Triticale
  international tests on, 40
  as man-made crop, 48
Tungro virus
  importance as pest, 44
  rice screening for resistance
     to, 25, 27

Underdeveloped countries, see
     Developing countries
United Kingdom, agricultural
     research in, 2
United States, agricultural
     research in, 2
Urea nitrogen, rice production
     and, 17

Vedalia beetle, insect pest con-
     trol by, 174
Vegetables, production in
     developing countries, 3
Velvet leaf, 56
Vigna mungo, crosses with
     V. radiata, 71
Vigna radiata, crosses with
     other Vigna spp., 71
Vigna umbellata, crosses with
     V. radiata, 71
Viruses, in biological control
     of sawfly, 183
Vogtia malloi Pastrana, as
     plant predator, 182

Walnut aphids, biological con-
     trol of, 185
Water supply
  agricultural research and, 42
  effects on rice production,
     18–22
    international tests, 27
  management, 23, 30–32

Water supply (cont'd)
  stress from, 56–57
Wax, on rice leaf, drought resis-
     tance and, 30
Weather, crop yield dependence on,
     55–57
Weeds, biological control of, 176
Wheat(s)
  for bread, 39
  dwarf characters of, 5
  haploid selection in, 69
  international tests on, 40
  irrigation for, 42
  major diseases and pests of,
     45
  new varieties of, adoption
     rates, 37
  photorespiration in, 104
  replacement of other grains
     by, 48
  tolerance to high aluminum
     levels of, 46–47
  "tropical", 48
  world yield of, 115
  yield of, photosynthesis and,
     54
Woolly apple aphid, resistance to,
     166

Xylan, products from hydrolysis
     of, 88

Zinc, soil deficiency in rice
     production, 21